QUADRING FEN SCHOOLBOYS OF 52 YEARS AGO

Proudly posing for a school photograph 52 years ago are these pupils of Quadring Fen school. The picture is owned by Mr. R. Howard, of Bicker, and he feels it will bring back nostalgic memories to many former pupils. On the photograph are: back row (left to right): Fred Howard, unknown, Clifford Machin, George Hempsall, Cyril Berridge, Fred Green, Cyril Stanley; middle row: Joseph Smith, William Griffin, George Green, James Smith, Arthur Barnes, Charlie Armstrong; front row: Stanley Mayne, Herbert Houghton, Frank Smith, Arthur Nelson, Robert Howard, Fred Carrott. The teacher was Miss Mabel Rylott.

The earliest known photograph of Joseph Smith who, as the newspaper caption states, is on the left in the centre row.

(We have been unable to trace the original. This is reproduced, with permission, from the Lincolnshire Standard of October 1965. (The photograph itself would have been taken in 1913 when Joe Smith was nine years old.)

A Life in Lincolnshire Title

From Plough to College

LIFE IN LINCOLNSHIRE TITLES

SOLD – REMINISCENCES OF A LINCOLNSHIRE AUCTIONEER
MOST BRUTE AND BEASTLY SHIRE (At present O/P)
I DIDN'T HAVE CANCER FOR NOTHING
FROM THE FENS TO WESTMINSTER AND BACK – OR WHAT PRICE INDEPENDENCE
CRIMSON SKIES – THE AUTOBIOGRAPHY OF A TWENTIETH CENTURY ENGLISH WORKING MAN
GROWING UP DOWNHILL – A BOOK FOR CLAIRE
OUT OF GRIMSBY
LARKS IN THE LINCOLNSHIRE MARSH

FROM PLOUGH TO COLLEGE

The **Life in Lincolnshire** titles, of which this is the ninth and may perhaps be regarded by 'furriners' as the one most typical of life in Lincolnshire, are all (auto)biographical accounts of the lives of people who have spent a major part of their lives within the county. Since one of the great attributes of Lincolnshire is its diversity it would be difficult to select a 'typical' life and although commonly thought of as an agricultural county cultivation of the land is but one aspect of a richly varied economy.

Important though the land might be, the economy of Lincolnshire includes the manufacturing industries of Lincoln and Gainsborough; the fishing industry of Grimsby – at one time, well within living memory, the greatest fishing port in the world; the steelworks of Scunthorpe; farming in the wolds; and the farming, now principally arable, bulbs and flowers of the southern fens. For more details of these and other of our Lincolnshire titles see the front and back flaps or inside the covers. Our catalogue is available on request.

FROM PLOUGH TO COLLEGE

by

JOSEPH H. SMITH, M.SC.

19 93

RICHARD KAY
80 SLEAFORD ROAD • BOSTON • LINCOLNSHIRE • PE21 8EU

The right of Joseph H. Smith to be regarded as the author of this work is hereby asserted in accordance with the provisions of the Copyright, Designs, and Patents Act 1988.

All rights reserved. No part of this book may be reproduced, stored in a retrieval system, nor transmitted in any form nor by any means, electronic, mechanical, photocopying, recording nor otherwise without the permission in writing of the publisher.

© Joseph H. Smith 1993
ISBN 0 902662 23 6 – cased edition
ISBN 0 902662 24 4 – paperback edition

Set in Bookman 10 point type for the main text using an AppleMac DTP system.
Printed by The Echo Press,
Echo House, Jubilee Drive, Belton Park, Loughborough, Leicestershire. LE11 0XS

Contents

Chapter	Page
Foreword	ix
1. Moving Times	1
2. In Fields and School	9
3. The Loneliness of Farm Work	27
4. Code of the Horsemen	38
5. Harvest Conflicts	46
6. Your King and Country Need You — on the Land	56
7. The Lonely World of 'Coäd and Squad'	65
8. Soldiers on the Farm	77
9. Trouble with the Furrows	87
10. 'Worthy of His Hire'	94
11. No Social Life	103
12. Union Members Under Pressure	115
13. Workers Locked Out	128
14. The First Escape	138
15. Chauffeur/Gardener — Back to the Land!	148
16. Back to the Lecture Room	157
17. After Early Doubts — A Diploma	165
18. In Harmony with the University	176
19. Towards a University Degree	185
20. University Don	195

Illustrations and Appendices

Earliest known photograph of the author ....Frontispiece

Appendices 203
    Letters from Annie Pitkin 204-6
    A Union letter 207
    A Reference 208
    UCW Debating Society 209
    In 'cap-and-gown' 210
    A Masters Degree 211

Acknowledgements

So many people gave me tremendous help and encouragement during my early years, both in Lincolnshire and at college, that it would be impossible to try to acknowledge them individually. In itself this book will present to the reader the size of the debt that I owe and will perhaps, in some measure, convey my sincere thanks.

J. H. S., Boston, September 1993

Foreword

'From Plough to College' would not, today, be the catching press headline that it was when I was fortunate enough to make the change some sixty five years ago. This is one of the consequences of improvements in the educational provisions which now enable many more of the less privileged young people to train themselves for positions in society more to their liking and abilities.

Many readers of this account of my early years may find it extraordinary that there were so few opportunities for young people living in rural areas to train themselves in an endeavour to escape from a life they found so dreary and unrewarding. Few of their parents had the financial means to enable them to stay at school beyond the age laid down by law. Parents with large families had of necessity to put their eldest children into employment at the earliest opportunity and having done so had no wish to treat their other children more favourably.

Those responsible for providing education in rural areas, mainly farmers and tradesmen, made a distinction between the needs of their own children and those of their employees. Others who considered it necessary to send their own children to grammar schools, and in a few instances to colleges and universities, showed little willingness to charge local rates with the burden of assisting able children of working class parents to continue their education beyond the elementary schools. Many young farm workers before and after my school days must have felt, as I did, that they had been trapped into an occupation without a future and from which they would be lucky to escape.

As the third child in a family of ten I had to accept that my parents had to get me away from school at the earliest possible date especially as at the time – the early years of the First World War – rising prices of food and clothing created serious housekeeping problems for my mother as they did for the wives of most men whose wages had been fixed for the year. I have sought to show how upsetting it can be to find oneself in a situation from which only the very few had much hope of escaping.

After seven years of elementary education, followed by about nine years working on farms, I gained a scholarship which enabled me to spend one year at Kirton Agricultural Institute. After that I worked for about eight months, for an American firm which in the mid 1920s was introducing a cyanide product into this country for use against farm pests. This work came to an end during the General Strike in 1926 and after twelve weeks without work, and a further seven as a gardener, followed by another spell of unemployment, I found myself back on farms as a farm labourer. After another year I was fortunate in winning a scholarship to Ruskin College, Oxford. That was in 1927. From there I went in 1929 to the University College of Wales, Aberystwyth, to take an Honours degree in economics with agricultural economics.

I have restricted this account of my experiences to the thirty years ending with the gaining of my degree. I might add, however, that thirty-three of the thirty-five years between my graduation and retirement in 1968 have been spent in Universities, either as a member of advisory staff working with farmers, or as a lecturer.

After leaving Aberystwyth in 1933 I joined the staff on the Department of Agricultural Economics at Reading University. I was there for less than a year before being invited back to Aberystwyth to work under Professor A.W. Ashby, Head of the Department of Agricultural Economics. Professor Ashby had earlier set and marked the papers for the Scholarship which took me to Ruskin College. He had also been one of my main tutors at Aberystwyth. In 1946 he left Aberystwyth to become Director of the Economics Research Institute in Oxford and I decided to make a change.

I accepted a temporary appointment in the Ministry of Agriculture. After two years there I joined the staff of the Department of Agricultural Economics, University of Nottingham where I remained for two years and then moved to the Department of Political Economy, University of Aberdeen as Senior Lecturer. This was a full time teaching appointment and was more to my liking than the advisory work which had occupied the greater part of my time at Aberystwyth and Nottingham.

I have made a number of references to farm wages and other small items of expenditure. These I have expressed in the old coinage prevailing at the time since it was not possible to convert to decimal coinage some of the small sums quoted without upsetting, in a few instances, the exact nature of the problems being considered. Those who wish to convert these sums from the old to the new decimal coinage should remember that the old penny was equal to approximately £0.004 in the present money, that twelve old pennies were equal to one old shilling and to five new pence.*

J.H.S.
Boston, Lincs.

---

* It is a mistake however to regard a simple conversion from the £.s.d. (pounds, shillings and pence) of the early years of this century — or indeed even the late interwar years — to the decimal sterling currency (£.p.) of today as being meaningful.

When Old Age Pensions were first paid (on 1st January 1909) the weekly rate was five shillings (5/-) for both men and women who were entitled to it (not everyone) at age seventy. It may not have been easy but it was possible to survive on that sum which, by a simple conversion to today's decimal currency, would be equivalent to 25p. — on which, of course, it would not be possible to survive. The basic pension today (1993), for both men at sixty five and women at sixty, is £56.10. Although it is not equivalent to the 5/- pension of 1909 it is at least as realistically comparable as is the 25p.

In the nineteen twenties and thirties a wage — anyone's wage — converted to today's decimal equivalent misleads rather than informs. In 1939, immediately before the Second World War, a packet of 20 cigarettes (large size) cost 1/- or, for one very popular brand, $11^1/_2$d. (£0.05p. or a fraction less). A 2oz. bar of chocolate cost 2d. and it was possible to buy a new motor car (Austin Seven or Morris Eight) for less than £120.

Interest rates in the Post Office Savings' bank were $2^1/_2$%. A nice three or four bedroom house could be built for much less than £1,000 and anyone who had saved some thousands of pounds would have been regarded as wealthy and could have expected to retire in comfort or even luxury.

To the memory of my wife

Chapter 1.

MOVING TIMES

AT THE TURN OF THIS CENTURY wages and working conditions on farms in Lincolnshire were said to be good. They were generally better than during the last quarter of the previous century if not better than at any other previous period. And they were better than in some other neighbouring counties which, like Lincolnshire, had little alternative industrial employment. Nevertheless most farm workers had a very hard time in the years before the Second World War. In the early years of their married life my parents had to contend with conditions only marginally better than those which their parents had experienced.

My paternal grandparents reared a family of twelve on a weekly wage rarely more and often less than 12s. In some years they could not afford to rear a pig for bacon and it was considered a serious matter when a cottager did not have a pig to kill at Christmas. Without a pig there was no pork for special Christmas fare, and no home cured bacon for the rest of the year. When that happened farm workers, who could rarely afford fresh meat, had to manage with the small quantities of bacon they could purchase each week out of their low wages. For my paternal grandparents it meant a piece of bacon which provided grandfather's supper on Saturday evenings and a meal on Sundays for those living at home. During the winter when farmers and cottagers killed their pigs grandfather begged pig's bellies which grandmother cleaned and cooked for the family. Until several of the older members of the family had left school and could add to the family income they had little meat at other times of the year.

My maternal grandparents, with a family of three, had a much better time. This grandfather, as a yearly hired man, received a fat pig each Christmas as part of his wages. But as they had to provide board and lodgings for two horsemen on yearly engagements the bacon from one pig rarely lasted the year. They could, however, afford to supplement it with weekly supplies of fresh meat. My maternal grandmother was also able to add to the family income by doing occasional work on the farm. She also managed, by gleaning after the grain harvest, to collect a quantity of wheat which the local miller ground into flour for the family.

My father's cash wage in the years immediately before the Great War (First World War) varied between 15s and 16s per week. In addition he had a free house, milk, potatoes, a fat pig at Christmas, and 18 gallons of beer for the harvest. As a yearly hired man he did not, like day labourers, receive any extra earnings for the longer day worked during busy times of year. His average weekly earnings, including payments in kind, would be somewhere in the region of £1. The pig, usually an old sow, when fat and ready for killing, weighed about 30 stones (420 lb. or 191 Kg.); the milk was skimmed unless, as happened with some of his engagements, he had milk from one cow. Father always had his 18 gallons of beer, others who, like my maternal grandfather, were teetotallers received 15s instead. On any reasonable assessment this payment for the longer harvest day represented a very low rate for the heavier work which one had to do at that time of the year.

My parents had some difficult times in the years before we older children started earning. There was a period of just under eleven years when they had seven children under school leaving age. The seven years between 1904 and 1911 were particularly difficult for in addition to, or perhaps because of, frequent pregnancies mother was in poor health. At one time she was unable to do all the housework and many of the chores had to wait until the evenings when father or the hired man living with us would give a hand by looking after the children or doing some of the housework.

Before the introduction of statutory control of wages and hours the contract year for married men, hired for the year, commenced on April 6th. Those who changed their employer at the end of their contract year had to vacate their service cottages – the tied house – on the morning of the 6th. Prior to the 1920s a large proportion of the workers in tied cottages made frequent changes of employer, consequently there was little risk of a man, who wished to make a change of employer, failing to find another who had a farm cottage. For single men and domestic servants on yearly contracts the year commenced on May 21st and ended on May 14th of the following year. The week between May 14th and 21st was taken as a holiday during which hiring fairs were held in each of the market towns in the county.

All sorts of reasons could be given to explain why some workers make frequent changes of employer and difficulties in personal relationships between farmers and workers may have been the least important. Family illness or other domestic motive obliged some workers to make a move perhaps to another district but the most important reason in the minds of many was the urge to seek a change in the hope of getting a higher wage, a better house, or a means of getting into another district more to their liking.

Fewer changes were made by men, married and single, who lived away from the farms on which they worked as day labourers. After 1918 there was a significant reduction in the number employed on yearly contracts. With no fixed terminal day to a contract of service, workers had less reason for making a change. It removed from their minds the anxiety of having, at a specified date in the year, to consider whether to seek or be obliged to find a new employer. And when a contract could be terminated at any time of the year by giving a week's notice there was less inclination to make a change. The disappearance of yearly contracts which had, within a particular region, a common terminal date meant that those living in service cottages who were required, or wished, to make a change of employer had much greater difficulty in finding other farm employment that carried with it the offer of a cottage. There are, today, many vacant farm cottages which are either no longer required by farmers or, being located in remote areas, are not sought after by workers.

The introduction of statutory control of wages and conditions of employment also played an important part in reducing the frequency of yearly removals. The minimum rates of wages prescribed in Statutory Wages Order related to the week. This did not, however, eliminate the right of farmers and workers to enter into yearly contracts provided the conditions did not fall below those prescribed in the Orders. In areas where the normal wage tended to be the statutory minimum workers had fewer opportunities for improving their wages and other conditions by making a change of employer. They were also less willing to enter into yearly contracts of service.

My parents, during their married life, occupied twelve different farm houses or cottages. By the time I left home in 1927 they had made eleven changes. Not all of them involved a change of employer. In one period of nine years, soon after their marriage, they moved home twice while working for Mr John Thorlby who farmed at Bicker. The first move was from Bicker Fen to Leasingham, where Mr Thorlby had rented a farm, and the second in 1905 to Broadwater farm, Moulton Chapel, which Mr Thorlby had rented after terminating his tenancy of the farm in Leasingham. Father stayed for a short while after Mr Thorlby had given up the tenancy in April 1910 and then moved to a free house in Bicker owned by Mr Thorlby. We were not there for long as Father became the foreman on Red House Farm, Donington Wykes, tenanted by a Mr Young. In another period of nine years – 1912-1921 – they again moved twice without a change of employer. The longest time they spent in one cottage was thirteen years but, in contrast to that, in a period of about two years between the years 1909-1912 they moved four times, three being into service cottages: one was a move of less than a mile whereas another was a distance of over thirty miles..

Each move was, for the children, an adventure. There was the excitement of new home surroundings while going to a new school offered new experiences some of which we looked forward to with apprehension. At each new school one had to establish oneself within the new society of school children, to prove oneself in the games which children played. And for the boys it was more than likely that one became involved in fights to discover whether, after April 6th, the new batch of boys included a new head of the playground. For mothers removals were at best mixed blessings. There might be occasions when the poor condition or location of a cottage, or the relationship with neighbours or farmer were so uncomfortable that a move was welcomed by every member of the household despite the cost and inconvenience. Most moves meant, for wives, increased work and expense as well as uncertainty about the future. And the condition of a new home was as likely as not to be no better, perhaps worse, than the old one. Husbands might, at the time of taking fresh employment, obtain fairly accurate information about their prospective employers and about working conditions; they certainly made careful enquiries on matters directly affecting themselves. One could be fairly certain, however, that few returned home after such an interview with a prospective employer with detailed information on such matters as the condition and exact location of the house offered with the work. Unless a wife was strong willed, able to insist, husbands rarely got from farmers information of particular concern to wives and children. It was unusual for farmers to offer, or for workers to ask for, an opportunity to inspect cottages being offered with hired employment.

In the days before 1920 it would have been difficult for wives to visit cottages offered by farmers. Few of them had cycles and in most cases it was impossible to walk the distances involved since husbands liked to move some miles away from the place of their previous employment. I don't remember hearing of a farmer making arangements for wives of prospective employees to examine cottages on offer. Wives had little if any opportunity to comment on the condition and situation of the cottages they had to live and work in.

Having moved into a new home there was little hope that the owner would pay any attention to an occupant's complaints about its poor state, although a contented wife of an employee would surely have enhanced the chances of a man giving satisfaction as a worker. I am pretty certain that my maternal grandmother would not have allowed grandfather to take work on a farm unless she was sure the cottage offered was in a satisfactory condition — but this was unusual. She was very houseproud and having worked hard to make her home spick and span would not have been persuaded, except for some very important reason, to move to another unless it was better than the one already occupied. With a family of three it had been easier for them than for most farm workers to settle down on one farm for the greater part of their married life.

Farm wagons were difficult vehicles on which to load large quantities of household goods and removals often resulted in small, occasionally large, breakages. As a consequence of our frequent removals father became an expert in packing our belongings on farm wagons. The heavy boxes of crockery went into the bottom of the wagon and had to be protected from the heavy pieces of furniture placed immediately above them. Bedding and soft furnishings were used to prevent items of furniture from rubbing against each other. But no matter how well the packing was done, or what care had been taken to rope down the load, one expected some breakages. There was also the risk that when loading the wagon someone would put a heavy foot on a piece of fragile furniture. The poor condition of both farm and public roads in early spring added to the risks, which, in our case, became greater as the family expanded and the quantity of household goods became too much for one wagon. We had more than most farm workers, because, in addition to our large family, father had, as foreman, to provide accommodation for a number of single men hired for the year. On a windy flitting day there was considerable risk that some of our high load of furniture would slip off the vehicle. Fortunately we never suffered serious breakages but, understandably, removals caused a great worry for mother who had no reserves of cash to meet costs of breakages. We were not insured against losses or damage to household goods during removal, in fact I don't remember my parents having insurance of any kind. We were lucky not to have suffered losses through fires or storms and only minor damages during removals.

It was not possible to move to a new home without having to meet extra costs. One usually had to do internal decorating. If one moved into a cottage where previous occupants had rarely stayed for more than a year, and perhaps for shorter periods, one could be pretty certain that little attention had been paid to the condition of inside walls and woodwork. Farm workers willing to attend to inside decoration delayed doing so for some time after taking a fresh house, giving themselves time to find out whether the farmer, the farm, and the district suited them. Few spent money on inside decorations if it seemed likely they would be moving again at the end of the year.

If the house had been previously occupied by the same family for a number of years one could be fairly sure to find internal decorations in a reasonable condition. Wives of men who made infrequent changes of employer had every incentive to spend to the limit to make their homes pleasant and comfortable. If the farmer was a good employer and he owned the farm one could expect not only that there had been few changes of occupant, but that the cottages would be in a good state of repair and had been well cared for by the occupants. One gained a fairly good idea of a farmer's qualities as an employer and manager from the condition of the cottages on his farm. It was, however, a mistake to assume that if farm buildings were in a good state of repair the farm cottages would also be satisfactory. There was a good deal of truth in the common complaint that farmers showed greater concern for the housing of their livestock than for that of their workers. Some tenant farmers who had difficulty in persuading their landlords to attend to necessary repairs and improvements to houses and farm buildings concentrated whatever pressure they could employ on getting repairs and improvements done to farm houses in which they lived, and to outbuildings. The condition of farm cottages ranked lower in their order of importance.

Inadequate and unsatisfactory water supplies added to the difficulties of housewives. My parents never had piped water prior to 1934. They had to be content with rainwater or with water from wells. The latter was invariably hard and unfit for drinking. Rainwater collected in butts or underground cisterns was used by most cottagers for drinking and washing. The supply was adequate in most years though the quality was often suspect. During mid-summer one had to be careful: in years of drought we had to carry water from other sources which, on occasions were some distance from the house. From the time when I discarded frocks until 1924 I do not remember having a bath. Washing hands, arms, face, and neck was done with care. As school children we gave ourselves a good wash every morning, mother making sure we left home for school without what we called a tidemark under our chins. In the summer we might pull off our shirts in the evening for a good wash. Feet we washed at irregular intervals, the frequency depending on what we had been doing in the daytime. The only time the rest of our bodies came into contact with water was when we got drenched by storms or went bathing in the large drains.

Lavatories – earth closets – were often badly sited, usually near the back door or at the bottom of the garden. In the summer they could be extremely noisome when full. One could be sure if one moved to a house which had previously had a series of short stay tenants one had taken a full lavatory. The responsibility for emptying them rested with occupants. It was never a pleasant task especially when, with yolks and buckets, the soil had to be carried some distance to a hole in the garden. One had to live for a long time with some nasty smells coming from these holes even though we covered them well with earth. At some cottages the lavatory was so close to the drinking water cistern that it presented a considerable risk of infection.

Chapter 2.

IN FIELDS AND SCHOOL

ONLY THREE OF THE COTTAGES which my parents occupied were in a poor state of repair and we stayed in two of these for only short periods. On these two occasions father, having decided to make a change of employer failed to secure other employment with farmers who had cottages to offer. As in each case we had to leave the house which went with the previous engagement he had no alternative but to take what accomodation was available and in both cases the houses he found were almost derelict. On another occasion we occupied, for thirteen years, a house which, though relatively new, was in a very poor condition. It showed every sign of having been jerry built. Plaster on all internal walls was crumbling when we moved into the house in 1921 and the farmer did nothing to remedy the defects during the time my parents lived there. It was impossible to make wallpaper stay on the walls. The farmer was too busy spending money on other things, including whisky, to give any attention to the condition of the cottages occupied by his workers. Had my parents been younger they would not have stayed so long. But having made so many moves in earlier years they had both reached a stage of wishing to have a more settled life.

Like many farm cottages this one was too small for our large family. Until some of us left home six boys had to sleep in one bedroom just large enough for two double beds. My sisters occupied one of the other two rooms. A common defect of farm cottages was the lack of accommodation for large families. It was, I suppose, expected that children would leave home soon after starting work. In most farm cottages kitchen work had to be done in the living rooms. If there was an outside washhouse it was rarely a convenient place in which to do the chores. Living room fireplaces, sometimes old and in a poor state, were inadequate to serve the needs of large families. The boiler at the side of the grate never provided sufficient hot water for all necessary daily needs. If one wanted a bath one had to use the copper provided for washing clothes.

Men, hired by the year, who lived in tied cottages, and who made a change of employer, did so on April 6th. Some might for special reasons move at the half year, on October 11th. Generally, however, only serious disagreement between farmers and workers caused men to move at that time of the year. With winter just ahead opportunities for gaining other employment with a good farmer were not promising and at that time of the year it was extremely difficult to find a farmer who could offer both work and a house. Those looking for men to occupy cottages in October would be suspected by workers of being bad employers. And farmers seemed suspicious of the qualities of workers seeking a change of employer in the autumn. There were, however, occasions when for the best of reasons a farmer had work and a cottage to offer at the half year.

The fear of eviction was always present in the minds of occupiers of tied cottages. Even those who had spent many years with the same farmer knew that, for a variety of reasons such as illness, old age, or a change of tenant of the farm, they might have to leave their cottage. Unless other accommodation could be found they might face eviction. For those who had to find employment at times other than the normal leaving dates this risk was considerable. Farmers had little difficulty in getting the necessary County Court Order enabling them to evict families from cottages. The most common reason given by farmers when applying for a Court Order was, and still is, that the particular cottage was required for the proper working of the farm, and was needed for the man hired to replace the one presently occupying it. Fortunately my parents were never evicted from any cottage but on one or two occasions they feared it might happen.

Towards the beginning of February married men, on yearly

engagements, had to make up their minds whether to accept an offer of re-engagement for another year if made to them by the farmer. If by Candlemas no offer had been made they knew it was time to search around for other employment. They scanned the local weekly newspapers for vacancies advertised by farmers and visited any offering work they considered suitable. As I have indicated earlier, a change of employer generally meant moving on to another parish since there was an understandable reluctance on the part of farmers to offer employment to men from adjoining farms. In their view it was in the best interest of workmen seeking a change to look for work in a fresh district. No farmer wished to be accused of stealing a neighbour's best workers and most would hesitate to engage men who had not been offered re-engagements by farmers in their immediate neighbourhood.

There were disadvantages in having to move into a district some distance from that of their current employer. When prospective employers and workers are strangers to each other each must depend on his personal assessment of the other. This is not easy during the short interviews which normally preceded an engagement. It was never easy for workers to get reliable information about farmers not personally known to them. Farmers were in a better position, opportunities to obtain information about workers seeking employment occurred during their weekly visits to the markets. Generally, however, engagements were entered into at the initial interview thus leaving neither farmer nor worker time to collect information on the qualities of the other. Unless workers were good judges of character, and capable interviewees, seeking new employers was bound to be an uncertain adventure.

I don't remember much about the changes father made during the early years of my life, I suppose they did not differ greatly from those I do recall. On the day of the removal we had to be up early, especially if a long journey was involved. One was lucky if the removal was not more than five or six miles. With these short distances there was sufficient time to take care in loading and unloading household goods. A horse drawn vehicle and man provided by the new employer came on the day of the removal. When longer distances were involved the man and his vehicle came on the evening preceeding flitting day. When this happened free stabling was provided by the farmer whose cottage was being vacated and the occupant of that cottage provided food and lodgings for the man.

When my father changed his employer in April 1921 he had arranged for his new boss to send two vehicles for the removal. However, only one was sent and it did not arrive until 10.00 a.m. on flitting day. By that time the farmer whose cottage we were leaving had arrived in an agitated state. Like father he suspected that our new employer had for some reason changed his mind about the engagement and had failed to warn father. We were relieved eventually to see the wagon arrive and to be out of the house before the new man arrived with his furniture. This illustrates the unsatisfactory method of negotiating an engagement at that time. Neither side had documentary evidence that a contract of service had been concluded. It was possible for a farmer to change his mind and yet take no action to let the worker know. Such action could have serious consequences for workers because of the difficulty of obtaining employment on other farms where cottages were available. Fortunately farmers had first to obtain a County Court Order before attempting an eviction of workers no longer in their employ.

On this occasion we had to travel some 15-16 miles to our new home and because of the late arrival of the wagon loading was done too hurriedly, with the result that the furniture was packed less securely than was father's custom. Several halts had to be made during the journey to adjust the load and re-rope it. Even so the furniture was hanging dangerously to one side by the time it arrived at the end of its journey. We had all, with the exception of father, arrived at our new home some hours before the wagon. Mother and the younger members of the family did most of the journey by train, this meant hiring transport for the three miles to the nearest railway station, then travelling by train some six miles before changing trains and travelling to within a quarter of a mile of our new home. Other members of the family did the journey by cycle. We had no food until the furniture arrived in the early evening. When it came there was a rush to get things into the house before it got too dark and in the hurry some pieces of furniture got slightly damaged. On an earlier occasion when we moved a distance of over 30 miles the journey took two days. This allowed plenty of time to load the things carefully. At that time the family was much smaller, we had fewer goods and loading them on a wagon was much easier. That removal was accomplished without any breakages or losses.

Apart from the inevitable cost in time long journeys had a further disadvantage if one had to pass several public houses. The number of calls made depended largely on the wishes of the man in charge of the horses. One had to offer him a drink; if he liked his beer and did not insist on paying his round then one hoped there would be few public houses to pass. I cannot remember father finishing a flitting day the worse for drink but we did hear of men arriving at their new homes incapable of unloading their goods.

At a removal every member of the family who could had to assist. So long as the work proceeded smoothly we small children enjoyed carrying things out to the wagon. The man who came with the wagon never took charge of the loading, thus avoiding any responsibility if breakages occurred. When difficulties in loading happened there was bound to be trouble for some of us youngsters before the end of the day. We had to watch carefully what we did. There was never time for proper meals and by the time we arrived at our new home everybody was in need of a hot meal. Younger members of the family, hungry and tired, had to be comforted. If the day had been stormy beds and bed linen, inadequately protected during the journey, had to be dried out before beds could be used. Parents, if not children, were always pleased to see the end of flitting day.

If the journeys were not too long they could be pleasant for older children who had few opportunities to travel from home. We had to walk but that was not a hardship on short distances. In

April there was plenty of activity in the fields and passing through villages and towns was exciting. For children who, like ourselves, rarely lived within easy reach of villages it was a special treat to gaze in shop windows even though we did not have money to spend on sweets. Longer journeys were pleasant enough for the first few miles. But when smaller children became tired older ones had the difficult task of urging them on. In the early days father managed to leave space on the wagon tailboard for mother and small children but as the family and the quantity of goods increased this became difficult and finally impossible. The pram with an infant in it was roped to the back of the wagon. When that became impossible those old enough took turns pushing it. I cannot imagine what townsfolk thought of us as we passed through on our way to a new farm. One could rarely use railways, long distance between railway stations and farm cottages often left us with no alternative but to walk. If flitting day fell on a market day some members of the family might travel part of the way by carrier's cart but that was only possible if loading the wagon had been completed before the carrier passed on his way to market.

Many parents who at the end of flitting day said 'never again' nevertheless occasionally found themselves on the move again at the end of another year. Some seemed to find frequent changes of farm and district worth the cost and inconvenience. Socially there were many disadvantages, especially for wives and children. Families never became members of a stable rural community, they had little opportunity to become members of rural institutions and associations and never spent sufficient time in any one place to become known. Many spent the greater part of their lives as strangers to all but the few families in their immediate neighbourhoods. It could hardly be otherwise for those constantly on the move from one isolated farm to another. One might live down a farm road, a long distance from a metalled public road. At best neighbours might live one or two fields away. As it was unusual for us to live within speaking distances of other families, mother had little chance of a chat over the garden fence.

Changes of farm generally meant a change of school for the

children. I don't remember the first one I attended. I had just started there when we moved in 1909. That was one of the two occasions, mentioned earlier, when father failed to get an engagement as farm foreman and a farm cottage. This particular move occurred outside the normal April date with the result that there were few jobs on offer. Fortunately the farmer who had rented the farm up to a few months before father left offered him a cottage in a village. It was a free house not attached to any offer of work. It was the only time we lived near a school and could go home for our mid-day meals. but we only stayed there for a very short time as Father soon obtained a position as foreman which took us to The Red House* farm at Donington Wykes. This was in 1910 and the farmer was a Mr Philips who lived on another farm.

This house was about the same distance from two schools, Wigtoft and Quadring Eaudyke, neither in the same parish as the farm. After spending a few weeks at one we had to transfer to the other because the number of pupils attending the first was too large for the available accommodation. While living there I first heard a gramophone, it played barrel shaped records. I also remember there was at the back of the house a large area surfaced with bricks, at least it seened large to us children in the summer of 1910 when we had to spend our Saturdays digging grass out from between the bricks. In later years Guy Fawkes Day always reminded me of our stay at this farm. Like most country children

---

* The Red House farm had been occupied in the late nineteenth century by a family named Swift. A daughter of this family, Sarah, became Matron of Guy's Hospital, London, in 1909, where there was, for many years, a gynaecological (diseases of women) ward named after her. The wing of the hospital containing this ward was destroyed in the 'Blitz' but a temporary replacement still carried the same name. Despite the changes of major rebuilding and alterations since the war there is still a ward named after her as Sarah ward, and a nearby block of staff flats is named Sarah Swift House.

She is perhaps best known as the founder of what is now the Royal College of Nursing.

15.

we could not afford fireworks or masks and we had to be content with scooping the middle out of a mangold or swede, making holes to serve as eyes, nose and mouth. With a candle inside we placed it on a thatch peg and went the rounds of neighbours. It was never a very worthwhile money earning exercise, few farm labourer's wives had pennies to give away.

This farmhouse was large. Some of the rooms were furnished and set aside for the farmer's son who came from time to time to overlook the farm. He often came at short notice and while there involved mother in a lot of extra work. My parents had not appreciated the amount of work caused by his occupation of these rooms. Mother was not in good health and the duties were more than she could accomplish to her satisfaction. Consequently father had, regretfully, to move again in April 1911. Our next farm, Sykemouth, Kirton Holme, was in the same area but in another parish and about a mile from a school. This was, for a number of reasons, an unfortunate move and we stayed for only six months. The farmhouse which we occupied was in good condition, and for mother the move had the advantage of being about four miles from the market town of Boston. For the first time for a number of years it was possible for her to meet her mother on market days.

The farmer, Squire Young of Swineshead Abbey, was not an easy person to work for. He had a private telephone which connected our farmhouse to the Abbey and this added to other difficulties experienced by both my parents. Mother had to be about the house during the daytime in case the phone rang and father found himself tied to the house in the evenings for the same reason. When it rang in the daytime mother had to get father in from farmstead or fields. Regardless of what she was doing, washing children, preparing meals or whatever, no matter how important it was, the farmer seemed to think it could be left while she went in search of father. With small children in the house it was never convenient to leave them unless an older member of the family was there. And the farmer got annoyed if he was kept waiting too long. Although these calls occurred infrequently my parents found the telephone a nuisance.

The summer of 1911 was glorious, the growing season had been good and the corn harvest was early. Two groups of Irishmen came to help us with the harvest and they 'kipped' in the barn. For reasons unknown to me the two groups soon began to quarrel and one left before the end of the harvest. This seriously upset harvest work by making it difficult to operate two adequately sized teams of harvesters. Father was unwilling to try and persuade both groups of Irishmen to stay, fearing serious injuries might result from the quarrels. Pensioners and boys too small for many of the tasks had to assist. It was not possible to ensure that the work progressed to the satisfaction of men, working at piece work rates, who wished to get ahead with the work while the good weather lasted. It was in their interest to have a short harvest since a long one usually meant low average weekly earnings and fewer opportunities for earning good money later by lifting potatoes. Delays also increased the risk of a break in the weather before the harvest was completed.

For the first time my eldest brother, not yet eleven, had to work and on the odd occasion I also assisted although only in the middle of my eighth year. My brother had to lead horses and I did so once when the distance between field and farmstead required an extra vehicle to keep the pieceworkers in the field and at the rick fully employed. When not leading horses my brother had to tend pigs on the stubbles. Water for the pigs had to be carried from a drain and as the task was too much for one boy I was directed by father to help. Normally a lad of eleven could be expected to tend the pigs without difficulty but in the special circumstances of the very hot dry summer it was impossible for one lad to carry enough water to satisfy the pigs. There was little water in the drain and the steep banks, boarded at the bottom, made the task especially difficult for two small boys.

On one particularly hot day we failed to carry sufficient water and the pigs wasted a great deal by getting into the troughs to cool themselves. The pigs decided to go into the drain and fell over the boarded side into the mud and water. We were scared and ran to father for help. Harvest work had to stop while the men came and drove the pigs to a point in the drain where a side ditch entered. The pigs were driven into this ditch and climbed out. This interruption annoyed the men since they were not paid for the time lost from their harvest work. It may have added to the tension which developed between father and the farmer following the trouble with the Irishmen.

Father was also irritated by the small wage paid to my brother for working that summer. When father was asked what he wanted for 'the boy' the farmer's tone indicated his annoyance at my brother's failure earlier in the week to stop the pigs from getting into the drain. My father told him to pay the boy what he thought he was worth. I suppose the farmer did just that for he gave him 3d. per day. I received nothing because I had been directed to the work by father without consulting the farmer. No fortunes came our way that summer.

King George V was crowned that year, an event celebrated in most villages and hamlets by giving Coronation Mugs and teas to school children. We received neither. I suppose each school ordered just sufficient mugs for the number of children on their registers at the time of ordering. Few school managers exhibited much sensitivity or generosity in such matters. If as a result of the yearly flit of farm families a school found itself with more pupils than mugs some had to go without. There were no spare mugs for us at our school. I don't know whether we received invitations to the school tea party. If so our parents must have decided not to let us go. Perhaps they thought it would be too upsetting for us to go to the school tea and not receive mugs. It was a great disappointment, our celebrations consisted of going to friends for the evening. Living in flat open country it was possible to see, on our way home, the sky full of colour from firework displays.

As children we never had weekly pocket money, family income would not allow it. The only times when mother gave us money were when we went to Sunday School treats or to local village sports. Then we received 3d. Those who had any luck in children's races had another penny or two to spend but I was never one to enter competitions.

It was unfortunate that we, like most children of farm workers, never had regular weekly pocket money. Given in moderation it trains children to appreciate the importance of wise spending. Children living in towns and villages might not have had weekly pocket money but they often did some of the family shopping which taught them something about the business of wise spending. We had no idea of the real value of money, no experience of shopping around for bargains. On the odd occasion when we had pennies to spend they left our hands quickly. In our excitement we never gave ourselves time to hunt for 'best buys'. Not that there were many bargains to be had on the sweet stall at a Sunday School treat. I may be presuming too much in thinking that children in my young days who did have pocket money hunted around for bargains, but I am sure that the modern child does. They are certainly more cute than I was at their age. Mother never encouraged us to spend money gained by doing odd jobs for people, it was put in our money boxes.

An example of my inexperience in shopping occurred in the summer of 1911. My eldest brother and I took, without permission, a shilling from mother's purse and my brother directed me to spend 6d. at a local shop not far from where we were tending pigs on the stubbles. I cannot remember whether he told me what to buy but I returned with twenty-four Barratt and Co's 'Lucky Packets'. I doubt if any town child at that time who had 6d. to spend would have restricted his purchases to 'Lucky Packets'. We ate the sweets and played with the small toys and charms. In the evening when doing repairs to my brother's trousers mother found several slips of paper in his pockets each with Barratt and Co on them. Being inquisitive she went through his jacket pockets and found the sixpence that we had hoped to spend on the following day. It did not take mother long to discover that a shilling was missing from her purse. Needless to say suitable punishment was administered and we lost the pleasure of having another twenty-four 'Lucky Packets'.

Since few children of our acqaintance received weekly pocket money we never thought ourselves unfairly treated. Indeed we had never heard of the term 'weekly pocket money'. We thought ourselves fortunate as mother gave us three or four boiled sweets most mornings to eat on our way to school. Few of our school companions had that pleasure. After leaving school and earning money we were given money each week and within three to four years handled all our earnings. We paid mother for our lodgings and bought our own clothes and other personal requirements. We had the same freedom over our money as the hired men who lodged with us. This soon taught us that our weekly wages would buy little more than the bare necessities.

Having decided, in the summer of 1911, to make a change of employer at the half year – October 11th – father sought for other work and a cottage. He failed to find a farm service cottage and the only one to let at a rent he could afford was in a very poor state. Several times during the winter of 1911-12 we got up in the mornings to find snow in some of the rooms. But this was not the worst of our difficulties that winter. Having broken his contract by leaving in October father lost the fat pig to which he would have been entitled at Christmas. This loss was a serious matter. A 'thirty stone' fat pig, which was part of father's wage, provided a substantial part of our yearly consumption of meat. My parents could not face the prospect of having to spend a whole year without any home cured bacon. They would have been obliged to do so but for the kindness of a friendly farmer who bought a fat pig and allowed them to pay for it in weekly instalments. This was, so far as I am aware, the only time when mother did not stick to the rule of buying only what she could pay for in cash. The pig was too small to provide sufficient bacon for us for a whole year and in the following April father obtained a situation as farm foreman with Mr G. Mowbray which involved providing board and lodgings for three single men. We became very dependent on bought bacon before Christmas 1912, when we again had a fat pig of thirty stones to kill. Most of this farm was situated in the parish of Gosberton, in the area known as 'Hundred Fen'. It was approached by a road which went over Meslam Bridge and joined the (now) B1177 half way between Billingboro and Pointon. There were, however, a few fields which were situated in the parish of Sempringham. When the South Forty Foot drain was cut it had separated these fields from the rest of Sempringham. Father stayed for nine years with Mr Mowbray, the first five as foreman and shepherd. He then moved to an adjoining farm rented by the same farmer. The duties of shepherd were taken over by another man and father, as foreman, overlooked the work on both farms. The move in 1912 took me to my seventh and last school. There I received four of my seven years of formal elementary education. Having attended so many schools I have no doubts about the soundness of the argument that changes of school are detrimental to a child's education. Much later, as a member of the City of Aberdeen Education Committee, I listened to councillors asserting that the upset to a child's education caused by being moved from one school to another was equal to the loss of six months training. I appreciated that elected members of the committee had, under pressure from parents, to resist frequent changes of areas served by individual schools. My response, however, was that, if councillors were right, then I had had less than five years of effective education. Many factors, including the individual child and the schools, have to be taken into account. Timid children do need time to overcome the upset of a change of school. Others with plenty of self-confidence are less disturbed, indeed for some a change of school may be educationally beneficial if it dampens down some of their cocksureness. One would expect most children to benefit from being moved from small ill-equipped schools in isolated areas which failed to attract good teachers. I have no reason for thinking that any of the seven schools I attended lacked good teachers but had I spent all my school days at one school my teachers would, perhaps, have been in a better position to judge my future potential, although that would, perhaps, have been of little benefit to me since my parents could not afford to let me stay at school beyond the earliest leaving date. They needed the extra income I could take home.

Some of the schools I attended were deficient in a number of respects. My last school was in an isolated part of the county and the authorities provided little money for books and essential equipment. Despite these deficiencies I managed to reach a standard of education which enabled me to leave school soon after my 12th birthday. I cannot say that I disliked my school days nor do I remember any member of our family not wishing to attend regularly. Weather conditions had to be very bad to keep us at home. We certainly received no encouragement from mother to stay at home and she could have made good use of our time. Children in urban areas, particularly those living in large towns and cities, could travel to school by public transport. Even in smaller towns where transport was not available walking to school in severe weather imposed less hardship on children than that experienced by those attending schools in remote country areas. Only one of my seven schools was less than a mile from where we lived but for most we had to walk much further. Open country gave us no protection from winter winds and storms. Except for one short period in 1909-1910 we had to take packed lunches consisting of bread and dripping and either a piece of cake or a jam pasty. Occasionally we had bread and butter with jam in place of bread and dripping. Most of our home made jam consisted of brambles and apples, the former we collected from the hedgerows. In time I grew tired of bramble-and-apple jam and have avoided it as often as I can since leaving home.

There was no special room in any of the schools where children could eat their sandwiches. In very bad weather we might be allowed to stay in one of the classrooms until we had eaten our food but teachers preferred to get us out of the rooms during the breaks. Badly ventilated rooms crowded with children soon became stuffy, especially when we arrived at school in damp clothes. Classrooms could be very cold during the winter. One open fire in each room was inadequate especially on the coldest days. Low temperatures made it difficult for us to concentrate on our lessons. In such conditions there was a good deal of shuffling of feet and much rubbing of hands. Provision in schools for dealing with wet clothes may have improved since my school days when no facilities existed. There was no heat in cloakroom porches and during the very severe frosty weather children could expect to find their overcoats stiff at the end of the day. Any attempt by teachers to dry stockings and other clothes had to be done in front of open fires, at the expense of lowering room temperatures and adding to the smells of overcrowded, badly ventilated rooms.

None of the schools I attended had a piped water supply. In most a can of water from a well or soft water cistern was provided in cloakroom porches for washing hands. We rarely used it and teachers did not bother to remind us of the importance of washing hands after going to the toilets. Before I left my last school a small water filter was placed in the girls cloakroom. I suspect it had been provided to supply teachers with filtered water for making tea during the mid-day break. It was not considered necessary to provide pupils with clean drinking water.

It was not possible in sparsely populated areas for schools to organise classes in narrow age groups. At my last school we had three teachers, one in charge of all pupils under seven, another taking the age groups between seven and nine or ten and the head mistress taking all the older children. The size of classes and difference in ages within each class meant teachers could not give adequate attention to the needs of each pupil. We had to work on our own for much of the time and since education often seemed uninteresting lack of teacher supervision resulted in children wasting a great deal of their time. Teachers had neither the time nor resources to keep mentally active pupils fully occupied. I found that able ones who demanded attention were directed to help less able children. By the time I had reached the age of eleven I was well ahead of other children of my age and spent much of my time in my last year at school helping other pupils with their work. Perhaps in helping them I gained more than I would have done had I been left to work on my own.

In larger schools each child had a better chance of getting the attention it needed. These schools had more and better equipment and a vastly better supply of books. At my last school we had few books that held the interest of country children. A small number of Dicken's *Oliver Twist* were provided a year or so before I left. The print was so small and tightly packed that it was difficult, especially on dull days, to read the story with ease. I like to think this and not the story put me off reading it. Those responsible for the purchase of school books failed to appreciate that the quality of paper and print have an important influence on children's liking for the printed word. Failure to provide rural schools with a supply of good story books or to establish public libraries in villages indicated an unwillingness to encourage working class families to improve their education and their ability to participate in public discussions on matters of interest to themselves and their communities. In my youth there was a general lack of interest by rural people in reading; we got little encouragement from our parents to read books and magazines. If more free time had been available to parents, perhaps they would have wished to read and have been more insistent that schools should have an adequate suppply of good books. Few rural communities in those days obliged their local councillors to provide village libraries.

When not at school we had plenty of work to do in or about the house and farmstead. Mother or father found work for us as soon as we arrived home from school. Dawdling on the way home did not help us to avoid any of the usual chores, indeed it often meant extra duties. There was never any suggestion that we should settle down in a chair and read books. In situations where children have no time for leisure pursuits there is little incentive to stay in the house. If we had a free moment we tried to get outside before mother thought up some other job to be done. We never went around looking for work. In such situations children soon lose any desire they may have had to read and learn about the world outside their own locality.

Children are naturally inquisitive and parents with time and a desire must find it easy to interest them in reading books which satisfy their curiosity. This in turn makes the task of teachers easier; both they and their pupils get a greater personal satisfaction in such situations. In my school days it would have been difficult to persuade most country parents to encourage their children to read books, and perhaps few had little idea of the kind of books that would prove most helpful to young people. Farm workers as well as many small farmers did not consider children needed 'book learning'. It would have been too optimistic to assume that, without the support of parents, any encouragement given by teachers would have resulted in increased reading in the home. Any special attempt in that direction would, in my young days, have met with strong opposition from parents wishing their children to help with work in the house or on the farm. And it must be admitted that for most children at that time it was more exciting to spend idle time playing or with farm animals

I cannot speak for all rural areas but, in South Lincolnshire, there was. and I believe still is, a great need for children to extend their vocabulary and improve their speech. It is, however, not certain that more time given by teachers to improving children's speech would result in an appreciable change. The influence of home and companions outside school classrooms must make the task of teachers extremely difficult.

Weekly reading coming into our homes in the years before 1920, in addition to two weekly local newspapers, was *Horner's Stories* (later named *Horner's Weekly*), *Family Journal*, and *Chips*. The two weeklies represented Father's reading before the first world war. After the war started our employer passed his daily paper to Father; this kept us more closely in touch with news of the war. Mother's reading was sporadic depending on the time she could spare from house duties. When all the normal daily chores had been completed there was always plenty of sewing and mending to be done. Her time for reading was scarce. As I grew older I often wondered why she bought weekly magazines.

Local weekly newspapers were important to farm workers in the spring. At that time of the year men living in tied cottages scanned the vacancies advertised for farm workers. Even those not seriously thinking of making a change studied the jobs offered, tempted by the possibility that there might be one which they would like to have. Normally papers and magazines came to our house on Saturday with the grocer's cart When, however, father had decided to make a change of employer he went to the nearest village to get the local papers as soon as they were published.

Anyone wishing to try for the best jobs on offer had to be away after them as soon as they appeared in the press.

I cannot remember when I began to take a continuing interest in reading. It was some time during the first world war. At the time there was a serial story called 'Red Cross Nancy' running in *Horner's Stories* which attracted my attention. This story took full advantage of our hatred of Germans which was at its peak. I soon discovered other stories in this weekly and in the *Family Journal*. When these failed to satisfy my needs I started to buy *Chips*. As I had by then left school one might think I was of an age for something more adult but lack of reading matter carried on the grocer's cart decided my choice. Another important factor was my ignorance of the wide range of good class magazines for boys that could be bought for a few pence. Had we lived in a village or town, near a bookshop or library, I might have selected something more suited to my needs. As it was it gave me and other members of the family, including Father, weekly reading matter for several years. The magazine stories may have been trivial but they served the useful purpose of getting me into the habit of reading. Schools never achieved that, the opportunites had been too irregular, too short in duration, and conditions in class rooms never conducive to quiet sustained reading. One might start but rarely finish a story.

In my last school I found one teacher keen to help pupils to extend their education beyond the narrow range of elementary school subjects. She sought to encourage a few children, including myself, to study French and shorthand. She gave, outside school hours, short periods of instruction once or twice a week. I never knew why these classes were discontinued after a few weeks. Perhaps the school cleaner objected because they interfered with her work or maybe the teacher was persuaded by her colleagues that she was wasting her time. It is also possible that the authorities raised objections. I am certain parents gave no encouragement for the instruction to continue, most, like my own, wanted their children at home to do various tasks about the house or farm.

Chapter 3.

THE LONELINESS OF FARM WORK

AS I HAVE ALREADY MENTIONED we always had plenty of work to do before and after school hours. In the mornings we had boots to polish and this was not an easy task during winter when boots were always wet; the 'blacking' used demanded a lot of brushing in order to get any sort of a polish. Other tasks included peeling potatoes and we needed a good half bucket each day. Rabbits and other pets had to be fed. As we grew older we looked around for opportunites to earn a little for our 'money tins'. We caught moles and rats. Mole skins realised between 9d and 1s each, depending on their quality and the farmer paid us 2d for each rat killed. We never managed to keep the rats down to reasonable numbers and the farmer had, at fairly regular intervals, to employ a professional rat catcher who came for the day with his dogs, ferrets and gun. We were not popular with this man because we took work from him.

From time to time the farmer required us to do a variety of simple tasks, often for no monetary reward. When he gave us a lift in his pony cart on our way home from school, we expected to be asked to do some kind of task. Most often we had to feed his hens and collect the eggs, a task which took up more time than that saved by getting a lift home. We never dared to ask for payment for fear of causing difficulties for father. There had been difficulty with a previous foreman whose sons tired of being at the farmer's beck and call. After having been used on a number of occasions and receiving no payment, they adopted a variety of stratagems to avoid getting lifts from school or coming into contact with the farmer when at play about their home. As a consequence the relationship between farmer and foreman was upset and the man left.

In the spring we had to assist the farmer's wife, in the evenings, with watering bedding plants. We carried the water and she did the watering. When the work had been completed the kitchen maid was told to give us a piece of cake and a glass of lemon drink. Each spring we assisted at rook shooting. Farmers in the district organised shooting parties in the evenings when the birds had returned to the spinneys. When it was the turn of the spinneys on our farm we went to carry the birds. For this we received six rooks which made a nice pie but hardly enough for all the family and the men lodging with us. I cannot remember being a beater during pheasant and partridge shoots but I often assisted, with other boys, at hare coursing during the Christmas holidays. We walked the fields to raise the hares. On these occasions a picnic lunch was provided and we were each given generous portions of pork pie and hot tea.

As the foreman's wife, mother had to look after a small flock of hens near our house, these were additional to those housed at other parts of the farm. Cleaning out the poultry houses near our home was a regular Saturday task for one of the boys. All the income from the sale of eggs, butter and poultry went to the farmer's wife with no deductions from it for any expenses incurred in their production. Mother was allowed eggs each week as payment for her time spent on the hens in her care. The number was never stated precisely and was expected to vary with seasonal changes in production. During the hatching and moulting seasons, when weekly production was at a minimum we had to be content with cracked and soft shelled eggs. Each spring chicks had to be hatched and reared; pullets were needed to replace culled hens, and the rest, together with the cockerels, were fattened and sold as Christmas table poultry.

There were always plenty of Saturday tasks about the house or farm. Brick causeys had to be washed and weeded, firewood had to be chopped for the week-end. We had little free time for games; mother made sure we did not move far from the house in case she needed assistance. During school holidays, more particularly in the summer, older members of the family worked on the farm. There was little one could do during Christmas holidays but at

Easter time we were employed scaring crows off newly sown grain. Crow scaring was not a task that I liked. When one had to overlook two or more fields the work was tiring. Not many farmers provided 'clappers' but we managed to make our own. These consisted of three pieces of wood, the centre piece being longer than the other two and having a shaped handle: the three pieces were then loosely fastened together with a leather thong. By shaking the contraption up and down the two outer pieces hit the centre piece making a loud noise. This was supposed to frighten the birds but they soon settled on another part of the same field or on the next one. The task gave us plenty of exercise running from one part of a field to another, or from one field to another. Without the clapppers we had to shout, this gave our lungs plenty of exercise.

When the fields were some distance from roads or houses one soon felt lonely and forgotten. Only those who have had this experience can fully appreciate the story told by W. H. Hudson in *A Shepherd's Life*, of a boy scaring crows on the Wiltshire Downs. Hudson, out cycling one day, saw in the distance this lad running towards the road. The lad arrived at the road out of breath and Hudson, thinking that the lad wished to seek information got off his cycle. As nothing was said Hudson asked the lad what he wanted and in effect was told 'nothing'. After further questioning about why he had been in such a hurry to get to the side of the road the lad told him: 'Just to see you pass'. That story always reminds me of my own experience in 1916. Soon after leaving school, I was sent to work some distance from house or road. Having no one to speak to and with no human being in sight the work soon became tedious. Time dragged, I continually looked at my watch hoping dinner time would come quickly. The hour break for dinner included walking time between place of work and home. Loneliness had such a compelling influence that I started for home well before dinner time, hoping that, by loitering, my father would not know I had left work early. He had only a short distance to walk from the farmstead and instead of arriving home several minutes after him I was first at the dinner table. That meant trouble; I had to take a quick dinner and go back to work. One wonders whether there is today the same failure to appreciate the effect, on youngsters, of making the change from the companionship of other children at school to the loneliness of farm work. Because of the small number of workers employed on most farms they often spent long periods working on their own.

Summer holidays were timed to meet farmers' needs for schoolboy labour. The six to eight weeks holiday was divided into two periods to fit in with the grain and potato harvests. In some years unsatisfactory weather conditions in the spring or at harvest time upset the intentions of those who fixed school holidays. If, as happened in 1912, the corn harvest was prolonged, because of storms, it delayed work on lifting maincrop potatoes. Although it was possible in my school days to adjust holiday dates at short notice, farmers could not be sure of getting as much help from school children in bad harvests as in years when weather conditions were ideal.

Harvest tasks for boys varied. Local workers, doing harvest work at piece work rates, had the help of their wives and children. The whole family would be in the harvest field unless the wife for one reason or another was unable to assist. Children too young to help played around the stooks or minded the baby while the older ones occupied themselves with the various tasks associated with tying and stooking sheaves or with other harvest tasks. Children of hired men and of day labourers on farms where harvest work was done at day rates could usually get plenty of work during the summer holidays. Crops flattened by storms had to be raised to enable the reaper knives to cut the straw near the ground without cutting off the heads of grain. On farms where a couple or more reapers were used several boys might be required to deal with storm damaged crops.

When grain crops were carried from fields to farmsteads, boys led the horses between loading and unloading points. I never cared for horses due, I think, to having to lead horses when too small to have proper control over them. On large farms a number of young horses would be broken into farm work each spring. By harvest time they were ready to do shaft work, but difficult for boys of eleven or twelve to manage. On hot summer days when flies were troublesome they needed to be under the control of youths with some experience of handling horses. A small boy could suffer damage from the hoofs of a fractious young horse. On a number of occasions I had my toes trampled on. Once I lost control of a young colt irritated by flies; it ran away with a load of sheaves and finished up in a deep ditch. Few farmers allowed boys to ride on farm vehicles during harvest time. It was considered too dangerous. One was supposed to have better control of a horse when walking at its head. But walking continuously to and fro thoughout a long harvest day was very tiring. Another summer task was tending cattle grazing the roadsides, a common practice in our part of the county in the days when only horse drawn vehicles travelled country roads. With the introduction of modern fast moving motor vehicles farmers dare not risk turning cattle out to graze roadsides. What I liked about this work was the company of children living in cottages along the roads and the chats with casual passers by. I never knew how much I was paid for these holiday tasks. Father collected our earnings and handed them over to mother. At that time, especially in 1915, she was glad to have every penny she could get and every member of the family able to work during school holidays did so if employment was available. My eldest brother and sister had both, by then, left school but the extra money they brought into the home did little to offset the rapidly rising prices of food and clothing. Mother always vainly hoped the money earned during the holidays would set us up with boots and clothes for another year.

Many parents considered it an advantage to get boys' suits on the large size to make sure none were discarded until worn out by one or other member of the family. We, however, had strong objection to wearing clothes which did not fit reasonabley well and I don't remember being ill at ease in new suits. Mother avoided, as far as possible, handing clothes down from older to younger members of the family. She knew we disliked the practice and considered that by giving us new clothes we would take greater care of them. None of us grew so fast that we had to discard any before they were worn out. Towards the end of the summer holidays mother went to market with a large list of items to be purchased. She never took any of the family with her but had details of sizes of suits and boots required. She managed to return with suits that fitted reasonably well. If boots were a little too large the toes could be filled with sheep's wool and an old felt hat could be cut up to make socks for boots. It was rare for mother to send any back as unsuitable; when it was necessary the carrier was commissioned to visit the shop for an exchange.

Yards of calico, shirting, flannel, and other materials were bought for making into shirts, vests and other underclothing. Apart from stockings mother never bought clothes she could make. She never had idle moments during the summer months though unlike wives of day labourers she did not work in the harvest fields. Until I was old enough to do my own shopping I never wore ready made shirts and vests. At the end of the summer holidays we went back to school, boys in their new smelly corduroy suits and heavy hobnailed boots, girls in their new frocks and pinafores. I must confess that stiff new boots could be rather hard on the ankles for the first few days but we were always pleased to show off our new 'rigouts'.

By 1910 young single workers and some married ones had cycles. Most women, and many men with large families of small children, could not afford cycles and depended on the railways and carriers' carts when visiting local markets. At that time I saw few boys riding cycles; only one,. a farmer's son, came to school on his bike. Families living too far away from railways to be able to use them depended entirely on tradesmen's carts and local carriers for their weekly purchases of essential household requirements. Most carriers attended at least two market towns each week. They worked to a fairly regular timetable and it was rare for anyone to wait long before being picked up. Those living long distances from the carrier's route might on occasions have to run when the carrier was earlier than usual. Carriers, however, knew their customers and never passed without halting at places where, each week, they expected country people to be waiting outside with their lists of shopping. Families living outside villages and away from the normal routes of tradesmen's carts were often very dependent on carriers. Those with small children could not make regular visits to towns except during school holidays and then only if they had children old enough to be left in charge of infants.

Frequent halts had to be made by carriers on their way to and from markets. Passengers had to be picked up and put down, shopping lists had to be collected, goods had to be picked up and taken to market. The latter included butter, eggs and poultry. On the way home goods had to be delivered to shops and houses. Almost everyone in villages and on farms depended to a lesser or greater extent on the carrier for the movement of some goods and for travelling to and from markets. Small village tradesmen had their weekly lists of goods required from wholesalers and carriers could expect some to require them to buy and sell goods in the market.

Carriers' carts, more correctly called wagonettes, had seats, some unpadded boards, along the two sides and the back. People entered through an opening in the front. The space down the centre of the cart as well as its tailboard would be occupied by baskets and parcels. On occasions the cart would be so fully laden with goods that passengers had the greatest difficulty in finding places where they could sit in comfort. Small windows in the sides and the back of the cart did not open, on dull days it was dark inside and on wet ones very stuffy. Not all carriers put candles or oil lamps inside during journeys on dark winter mornings and evenings; most feared passengers might upset the candles or lamps and cause fires. Without lights the journeys could be very wearisome.

Travelling to and from markets at Christmas time could be particularly difficult and uncomfortable. Many housewives went to market at that time and came home laden with Christmas shopping. Carriers also had extra goods to take to and bring from markets. The journey home was worst because of the large number of parcels. Passengers were tensed up with fear that their Christmas toys might get broken. As they crowded into the cart for the return journey it became increasingly difficult for late arrivals at the point of departure to get past people and parcels to a seat. The first in seated themselves near the entrance unless it was very stormy or cold when they took seats at the back of the cart. Other busy days for carriers were at the New Year and during May hiring week. Single men and domestic servants who, for one reason or another changed their employers in May, arranged for carriers to collect 'boxes' (chests) containing their belongings. Carriers took these to their owners homes if along the route, if not, then to market and transferred them to other carts for the rest of the journey. This task was commonly done on the days of local hiring fairs when carriers could anticipate extra passengers going to the fairs. Passengers could expect to be tightly packed in the carts with knees rubbing against large wooden or tin chests.

Carriers arrived in towns about 11.00 a.m. and started their homeward journeys at about 2.30 p.m. On Fair Days they delayed the return journey a little but most of them liked to get away soon after 3.00 p.m. because of the distances to travel and the number of calls to be made on the journeys. Naturally young people wished to stay longer in town on Fair Days and some who failed to be at the place of departure on time either had to walk or find other means of getting home. By the early 1900s, however, most young men and women who had worked for two or more years had cycles and they were no longer dependent on carriers except for the removal of their 'boxes'. The time spent at markets was fully occupied by carriers. Produce had to be taken to the Butter Market and parcels to shops and warehouses. Shopping could take a long time if housewives insisted on naming shops from which goods had to be purchased. Some asked that their orders be given to a named assistant of a particular shop because 'he' or 'she' 'knows what I require'. Carriers preferred to shop where they pleased, it saved time and gave them a better opportunity for getting commission from shopkeepers. They soon became familiar with the individual requirements of customers especially of those who had used their services for several years but it was not always easy to learn what some customers wanted because they gave instructions in extremely vague terms.

Married men living on outlying farms made few visits to their market town, especially if they lived ten or more miles from it. A few attended the yearly hiring fair but most could not afford to lose a day's pay. Visits at other times had to be on Saturday evenings. Before the introduction of the weekly half day, work on most farms did not finish before 5.00 or 5.30 p.m. and unless one lived within three or four miles of a town one had very little time to shop — even when shops and stalls did not close before 10.00 p.m.

Horsemen, who normally finished work at 7.00 p.m, had little hope of reaching a town in time to do much hunting around for bargains. Some farmers, employing two or more horsemen, allowed each, in turn, to leave work on a restricted number of Saturdays, at the same time as the other workers. This, however, was not a general practice, most horsemen had either to rely on their parents to do their shopping or visit the market themselves after finishing work at 7.00 p.m. I remember there were occasions when those who lodged with us went to town without waiting for their evening meal. Even so they could not do the trip and get back by 9.30 p.m. which was, except on special occasions, the time father expected lodgers to be in the house. Horsemen on small farms had the greatest difficulty in leaving work early to do any shopping. Unless their employer was willing to finish off the stable work they could never get away before 7.00 p.m. during the winter period when horses had to be stabled for the night. For many horsemen, living a long distance from town or village, most Saturday evening trips were limited to a visit to the nearest public house. Shopping was a little easier during the summer months when horses were turned out to the pastures for the night.

The local village sports was for most men an event that could not be missed. These annual events appealed to rural people of all classes and ages. Wagons decorated with flowers and bunting took families from outlying farms. It was the one occasion when horsemen displayed their collection of horse brasses, when every effort was made to polish harnesses and to wash wagons. As school children we were taken from Quadring Fen school by wagons to the annual sports at Quadring. In 1912 the party included five from our family. First we went to the church for a service, then to the village school for a tea, and finally to the sports field. Later parents came, again in wagons, in time for the evening events. These consisted of the usual athletic events. I don't think that the sweets stalls did much trade, children of farm labourers had little money to spend. The beer tent fared better, though some of the men from farms never got nearer to the sports field than one of the village public houses.

We returned with our parents in the wagon which had brought them and younger children to the sports. There was the usual difficulty of getting people collected together for the journey home. The odd man, having had too much beer, would be quite merry, and might need some persuasion and help before he was safely in the wagon. Harassed mothers had the children as well as trying to keep the rest of their families near the wagons. This was not easy with several vehicles and people around. The men were much too busy gossiping to give a hand with the children. If any men wandered away it was important not to let the less sober men assist in the search, to do so involved the risk of having to search for them as well which might mean visiting several public houses. It was advisable to make certain that each person, man, woman, or child got into the wagon as soon as they arived at the vehicle for the journey home and allow none to get off until they had reached home. It was not an easy rule to enforce.

On the journey home we had a mixture of singing from both children and adults and crying from small, tired infants. Singing was accompanied by music from mouth organs and accordions. Disharmony was not confined to singers and musicians It was never easy, especially with a drunk in the company, for adults to keep tempers under control and to conduct arguments in a temperate language. Some women hoped that the singing would have a soporific effect on both crying children and argumentative husbands. Most often the argument was between wives wishing to get home, to get their infants to bed, and husbands wishing to have a stop for liquid refreshments. If the man in charge of the wagon was a moderate drinker and sympathetic to the wishes of the womenfolk he might refuse to make calls at any of the public houses on the way home. A determined man might however get out of the wagon as it passed his favourite pub. When that happened he had to find his own way home — not always an easy task.

Chapter 4.

CODE OF THE HORSEMEN

THE LENGTH OF THE WORKING WEEK at the present time is considerably shorter than when I started work in 1916. At that time general labourers worked a week of fifty seven hours in the summer and fifty one in the winter. There was no weekly half-day before 1918. From the beginning of April to the end of September, labourers worked each day from 6.30 a.m. to 5.30 p.m. and for the rest of the year they started work at 7.00 a.m. and finished at 5.00 p.m. for the day, half an hour earlier. They had half an hour for lunch at 9.00 a.m. and an hour for dinner at noon. Special workers, such as foremen, shepherds, cattlemen and horsemen, men hired for the year, worked what was known as the customary hours in their particular area. For cattlemen weekly hours depended on the amount of work that had to be done on Sundays. During the winter when the cattle were 'yarded' Sunday duties, morning and afternoon, might involve three or more hours. In the summer Sunday duties could be accomplished in less time. These weekend duties represented time worked over that of general labourers. Shepherds had to be on the alert day and night during the lambing season whereas at other times of the year their weekly hours might be in the region of sixty to sixty five, and often the same as day labourers. Weekly hours for foremen depended on the status and duties of each. In theory at least, they had to be available at all times of the day and night, but those of higher status, whose responsibilities were wholly supervisory, had greater freedom to control their hours of duty.

The total weekly hours worked by horsemen could be as high as seventy during the winter months when horses spent all their rest time in stables. From the middle of September to the end of April horsemen started their day at 4.30 a.m. and finished at 7.00 p.m. Some farmers might allow men to start their day at 5.00 a.m. but most considered that horses should have at least two hours feeding before leaving the stables for a day's work in the fields. It was also considered that stable work could not be properly completed in less time during the winter months. In the summer when horses grazed the pastures during the night horsemen might be allowed an extra half hour in bed and to turn their horses out to grass at 6.00 p.m. These concessions however, still left horsemen with a working week that was ten to twelve hours longer than that of a general labourer.

Horsemen had half-an-hour break for breakfast at 6.00 a.m. during the summer and 6.30 a.m. during the winter months. They normally had their lunch break at 10.00 a.m., and dinner at 3.00 p.m. During the winter period when horses spent all their rest periods in the stable horsemen had their tea at 7.00 p.m. During the summer months when horses spent the nights out on the pastures the evening meal might be taken at 6.00 p.m. Day labourers, when in charge of horses doing field work, had lunch and dinner breaks at the same time as horsemen.

Each evening the farmer or his foreman instructed the horsemen on the work planned for the following day. On most farms in our district they were required to harness the horses before going for their breakfast.

During harvest these men had the same meal times as day labourers, that is lunch at 9.00 a.m., dinner at noon, tea at 4.00 p.m. and supper at the end of the working day which might be 9.00 p.m. Like other workers they took their lunch to their place of work, often in fields where, in winter, it was difficult to find shelter from winds and storms. I have spent many cold, uncomfortable, meal breaks in open fields trying to find some shelter and warmth from my team of horses standing in harness and attached to whatever implement or vehicle they were pulling. No food or water was provided for them when at work except when working a long harvest day.

It was said that every good horseman was a thief, a stealer of grain and meal for his horses. I remember in 1911 father had the greatest difficulty in preventing horsemen breaking into the granary to steal oats. Ordinary locks never stopped a determined man from getting the extra grain he thought his horses needed. They could file keys to fit the lock commonly found on farms. The daily ration allowed by most farmers was 14lb of grain or meal per horse per day when wholly hand fed and half that quantity when they spent their rest periods on pastures. Although few horsemen considered 14lb sufficient when heavy work had to be done perhaps few actually broke into barns and granaries to add to the ration. There were other ways of getting a little more grain. When threshing was being done it was often possible to persuade the men carrying the grain from the thresher to the barn to put small quantities in the chaff house, making sure to cover it up with chaff or oat straw. This could be most easily done when threshing and chaff cutting were done as one operation.

When a man was under suspicion of stealing there might be a good deal of snooping around the chaff house by his employer or foreman when he was away from the stable. Any grain or meal found was removed. The incorrigible thief expected any locks he put on the corn bins to be tampered with as often as he tampered with those on doors of barns or granaries. Farmers could do little to stop the practice short of dismissing culprits and there was always the risk that the next hired man would not behave differently. Unless the quantities stolen were excessively large nothing was said and horsemen deemed it wisest not to comment if some stolen grain disappeared from corn bins and chaff houses. I doubt whether stealing was ever as bad as some farmers and horsemen claimed.

In the days when horsemen did a lot of road work, carrying produce to railway goods yards and to merchants' premises; when lime, fertilisers and coal had to be carried to farms, no horseman liked to be doing the work with horses not in prime condition. None liked to hear, from other horsemen, uncomplimentary remarks about the quality and condition of their team.

The use of drugs was treated more seriously by farmers. It was said that some men used drugs to encourage their teams to put on flesh, to give horses' coats a fine shine and to make them more lively on the roads. Horsemen liked their teams to be like prima donnas, temperamental and a little skittish when on road work. Although there was always a good deal of talk among horsemen about the various drugs which could be used, my impression at the time was that very few actually used them.

A well established code of conduct was observed by horsemen and others in charge of horses. A head horseman had first choice, from the stable of horses, for his team. Others took their choice in order of seniority. Any farmer wishing horsemen to take, as their team, horses other than those of their choice had to have very good reasons. One reason might be to provoke a horseman to hand in his notice before the end of his year. An employer might wish to be rid of a horseman but unless he had reasons which gave him a legal right to dismiss a yearly hired man some other means had to be found. If a man could be provoked into handing in his notice that served the farmer's purpose.

In the mornings the head horseman led the team out of the stable. If for any reason he and his team had to stay back the next senior horseman led the teams out. Teams going to and from work, or when doing cartage work, followed each other in strict order. Anyone who dared to break the rule could expect trouble, particularly if the head man was not only a good horseman but also a forceful character. Such men demanded to be given first offer of any road work, especially journeys to places of importance. When more than one team was involved the head horseman decided any halts. If he stopped at a pub other teams had to stop even though the men might not wish to have a drink. It was said that horses from some farms knew all the stopping places between farm and railway station or market town. These horses, it was claimed, needed a good deal of urging to pass the customary halts. It was also claimed that they could be relied on to take themselves and fuddled horsemen home without mishaps. That I think was just 'stable talk'.

Some farmers provided special harness for particular occasions such as agricultural shows and taking children to village sports. At one farm where we lived sets of brown leather harness were provided. These hung in our kitchen. I don't remember them being used during our short stay on that farm. In mother's view they were a great inconvenience, collecting dust and taking up wall space which she considered would have been more usefully occupied by flitches of bacon.

In addition to the week's holiday in May, horsemen had three days holiday at Christmas or at the New Year. The head horseman had the first choice of these two winter holiday periods, if he decided to take his holiday at Christmas the other horseman, if only two were employed, had his break at the New Year. If more than two were employed then a day labourer would be asked to assist with stable work during these holidays. On some farms the horsemen on duty had to do the extra work.

I don't know whether the contract year ended legally on May 14th or 21st, but have always assumed that it was the latter date and that the week in between, and the holidays taken at Christmas time, were intended to some extent to offset the long working week. The men regarded both periods as paid holidays. Occasionally men who accepted re-engagements for another year might be asked to continue working over 'May week'. Payment for this week was outside the yearly engagement. Hiring fairs were held in the local market towns on market days falling between May 14th and May 21st. In each town there was a particular street or place where men who wished to be hired, stood. Not all those there would be seeking 'a place' and it was not possible for a farmer to distinguish between those wishing to be hired and others there for a gossip. There was always plenty of opportunity for men to have some fun with the not so bright lads wishing to be hired, or with those who had spent too much time and money in the pubs. Those seeking engagements might demand conditions out of the ordinary but, as the week advanced, those not hired lowered their demands, became less particular about a farmer's reputation, about the location and condition of farms, or about the number and condition of horses. Those not hired by the end of the week had to seek casual work and might have to wait until the next year's fairs for an opportuniy to be hired as horsemen.

No one liked standing around waiting to be hired: on a rainy day it was most unpleasant and on a warm sunny day tiresome. There was always the urge to be away with one's pals, to the pubs or to join in the fun of the pleasure fair. On being hired men received a 'fastening penny' usually half-a-crown, which made the contract binding. After being hired some men spent the rest of the day in public houses. For them the journey home could be difficult, both in trying to find their cycles and in mounting them and cycling home. Fish and chip shops and other 'dining rooms' did a good trade. Country folk referred to restaurants and cafes as 'dining rooms'. The most popular mid-day meal by those who favoured these 'dining rooms' rather than fish and chip shops was sausage and mash. Few bothered to stay for a second course. After the meal they went either to one of the stalls for a bottle of pop or back to one of the pubs for something stronger. Afternoon teas, for those who still had money to spend, varied. Some had to be content with bread and butter and a pot of tea, others might manage a fruit tea and a few, perhaps, tinned salmon.

Farmers wishing to re-engage their horsemen for a further year appproached the men before Candlemas, February 2nd. Some men, however, might not be prepared to make a firm decision in advance of the May hirings, wishing, if not to test the market themselves, at least to know the market trend in wages. On being approached, a man, while intimating an interest in being re-engaged, might suggest a meeting at one of the local hiring fairs to finalise an agreement after more was known about the general level of wages being offered. If a farmer seemed determined to secure a re-engagement his horsemen might decide it was a strong indication that farmers expected wages to rise. On the other hand if an employer was reluctant to make offers in advance of the hiring fairs his horsemen became apprehensive about the prospects for the coming year. It might be that a farmer was dissatisfied and intended to make a change, or perhaps he expected wages to fall and in consequence delayed making an offer until nearer the end of the current year. Some horsemen delayed a firm acceptance of an offer made before the hiring fairs because they wished to avoid being asked to continue working during the holiday week. Most horsemen liked to be free to attend the hiring fairs in their area.

At the hiring fairs one learned a great deal about the reputation, or alleged reputation, of farmers, as farmers and as employers. It should be added that some of the views expressed by workers were influenced by personal animosity. Men annoyed at not being offered re-engagements could be less than fair in their criticisms but the discerning person, attending these fairs and listening quietly to conversations, could usually distinguish between what was true and what was false. Farmers also exchanged information on men seeking employment which could be less than fair to the men concerned.

Apart from the general information on farmers, horsemen wanted to know about the quality of the horses owned by those seeking men. They particularly wished to know whether a farmer did a lot of trading in horses, whether he bred large numbers for sale after breaking them in. On farms where a good deal of breeding was done head horsemen might find themselves losing their teams several times during the year. It was perhaps all right to be in charge of a stable where sales were restricted to newly broken-in horses or to those no longer able to stand up to the heavy work, but few horsemen liked to work for farmers who frequently sold the best horses from their stables. They preferred to be on farms where few sales were made and those of the worst, not the best horses. Having trained horses to their particular ways men had no wish to see them sold.

In the First World War farmers were obliged to sell a considerable number of horses to the military authorities. Some complained of having to sell a higher proportion than their neighbours. This often happened with farmers having stables of top quality horses. Other farmers complained of the low prices paid by the authorities, and, because of the shortage of horses, of having to pay much higher prices for replacements of poorer quality than those taken by the military authorities. The best class of horsemen tended to be on farms with the best horses and they had to put up with the inconvenience of losing proportionally larger numbers to the army. They complained that farmers did not offer sufficient resistance to the demands of the army, especially when the reduced number of horses on farms had to cope with an increased area of ploughed land.

Although dislike of a farmer, his farm, the district or of the lodgings provided, caused horsemen to make changes of employer, these may not have been the most important reasons why few young single men stayed more than one or two years on any farm. Most liked to use a change of farm to widen their experience of farming districts, of different systems of farming, and of farming practices. This was particularly so in the case of men hoping to progress from being horsemen to being foremen. For the restless type, those free to go where they pleased, the end of a contract year was good enough reason for making a change.

Chapter 5.

HARVEST CONFLICTS

FEW PRESENT DAY FARM WORKERS have any personal experiences of farming conditions before the First World War. Changes that have taked place during the last fifty years have not only improved methods of production and sources of power used but have also involved the transfer to outside agencies of a number of important tasks previously done by farmers and their workers. There have also been considerable improvements in the quality and yields of crops, and of livestock and their products. These changes have had an important influence on farm work and on the type of worker required. At the turn of the century farm tasks such as harvesting and threshing cereals involved teams of workers and a number of separate work processes which today have been organised into one single task requiring smaller teams of workers. There has been a tremendous reduction in the total number of people employed in British agriculture while at the same time the total quantum of output has been increased resulting in a very large growth in the output per person employed. In the past a greater proportion of the tasks required brawn rather than brains. Today mechanical aids have removed much of the heavy lifting from work, and removed the drudgery that earlier farm workers had to accept as a necessary part of it. Many of the skills required in my youth have gone, a different kind of craftsman is now required, one having a wide range of mechanical experience.

When the turning and cocking of hay was done by hand most farmers managed the hay harvest without employing extra labour but casual labour had to be employed to help with the grain and root harvests. The work in the hay and corn harvest was heavy, in a wet season it could be tedious and dirty. This was particularly so in the case of hay making. On farms where elevators were used to convey hay and sheaves from vehicles to ricks the work was less unpleasant. On most farms before 1914 reapers and self-binders were the only mechanical aids used to harvest hay and grain. Most grain crops were cut by sail reapers. Self-binders were used only for crops of standing corn free of excessive amounts of ground weeds, especially of bindweed. Using self-binders on crops badly damaged by storms or badly infested with bindweed caused endless holdups due to the machines getting clogged up with the tangled straw. For such crops farmers preferred to use reapers which allowed time for weeds to wither before tying and stooking. Sail reapers, more correctly called side-delivery reapers, deposited crops in bundles of sufficient size for tying into sheaves. In well standing crops these machines did a neat job, making the work of tying and stooking easy. But with crops damaged by storms and weeds it was often impossible for them to leave neat bundles with all the ears of grain lying in the same direction. In order to keep reapers and self-binders fully employed when conditions allowed, two teams of horses were needed for each machine. This allowed each team to have rest periods and to be fed and watered. In a normal season two horses would pull a reaper: three horses however might be required if the ground was soft and wet. Three had to be used for pulling self-binders. When petrol engines, attached to the machines, operated the cutting and tying mechanisms two horses could pull the 'binder'. Later tractors pulled the machines and eventually also operated the cutting and tying mechanisms as well.

In addition to the men in charge of horses at work and resting others had to be employed on a variety of tasks about the field. The knives of the machines had to be sharpened, patches of badly damaged crops had to be lifted so that the reaper knives could get under the crop, and the worst parts often had to be cut by scythes. When self-binders were used lads had to move sheaves at the corners of the uncut crop so that horses could move round without trampling grain out of sheaves. When sail reapers did the work men mowed off sharp corners as these developed, this allowed the teams to turn corners without trampling on the cut crop and, as I have previously mentioned, boys lifted up badly laid patches of crops so that reaper knives could cut the straw without loss of grain.

Meal breaks were staggered to ensure that work continued without interruption. Men went in groups of two or three to have their meals. The normal times for meal breaks had to be ignored and no one could take the normal break. Men employed on tying and stooking who had their wives helping them had few oppportunites for hot meals other than at weekends. In our case mother was able to send out to the field hot dinners and cans of hot tea for the afternoon meal. Those taking the first break for dinner had a reasonably hot meal but others who came later might find rations both short and cold. It was much the same at the 4.00 p.m. tea break. The quantity of food and tea mother sent out depended on the number of people it was meant for but it was never easy to judge just how much hungry men and boys needed. The lunches and cans of tea were well wrapped in an endeavour to keep them hot. Children carried the food out to the field which might at times be some distance from the farmstead. Although mother instructed us not to loiter there was a risk that we would find something which attracted our attention and caused our journey to take longer than it need have been.

On all the large farms where Father worked as foreman, tying and stooking was done by the piece and Irishmen came to help. Local married men worked in separate family groups. Others, including Irishmen, might work each on his own or in small groups depending on personal relationships. A single piece-work rate was negotiated for all the work of tying, stooking, and hand raking between the stooks. Each farmer preferred to fix one rate for all the tying and stooking to be done on his farm or group of farms if he had more than one. This avoided interruptions on work while new rates were negotiated as would have been necessary had each field been treated as a separate contract. Difference in the condition of crops, as between one part of a field and another, as well as between fields might seem to justify fixing more than one rate but both farmers and workers sought to avoid having a multiplicity of rates. I only remember one occasion when some men, not regular workers on the particular farm, left before the end of the harvest their reason being that the farmer would not negotiate a new rate for a crop which had been badly damaged by storm. The men argued that since the damage had been done after the initial rate had been fixed, they ought to have had an opportunity to negotiate a new one in circumstances where the work had changed materially from that anticipated when harvesting began.

Conflicts between regular and casual workers often arose when severe weather conditions affected harvest work and justified attempts by workers to renegotiate a new piece-work rate. Farmers in resisting these maintained that workers had in mind the uncertainties of the weather when agreeing the rates. This was rarely accepted by some men and certainly not by casual workers. Success in gaining adjustments was most likely where men, living in tied cottages, formed a relatively small proportion of the total number employed on the work. Regular workers had little choice but to observe the terms of an agreement. Their future employment on the particular farm and for some the occupation of a farm cottage obliged them to do the work even when the effect of bad weather on the crops had reduced weekly earnings well below what had been anticipated at the beginning of the harvest. In the case of casual workers other methods had to be employed by farmers wishing to hold them to the contract. Farmers could, and some did, limit weekly drawings by men on piece work. They sought to hold a large balance of earnings until the end of the harvest. Any man who left before the work had been completed risked losing some if not most of the balance of earnings held by the farmer. Some men did leave a farm during the early days of a harvest if the expectations for earnings were not to their liking but as the season advanced and the balances held by farmers increased they had little alternative but to stay and complete the work. There was also an understanding between farmers not to take workers from each other during the busy seasons. No farmer, however, could hold a man determined to leave; the man could become such a nuisance that anyone would be glad to give him his money and let him go.

The men never knew in advance which of the crops they would have to tie and stook and which they would only have to stook. The work was much more difficult if a crop had been badly damaged by storms or was badly affected by weeds, particularly cornbind. Where storm damage had been severe the straw might be so rotten that it was difficult to find any strong enough to use for binding the sheaves. Consequently a uniform piece-work rate was only possible when all workers accepted the method of allocating work. A commonly accepted rule was for each man or team, at the start of the harvest, to draw lots for the order of taking plots (called lands). Each land was approximately a chain in width indicated by furrow markings, and ran the length of the field or crop. As each man or team completed the work on one land he/they moved to the next land not being worked. The same rule applied when the men moved from one field to another, the first to go into a new field took the first land on that side of the field nearest the gate. The intention was to eliminate any selection, by men, of particular lands in a field. The system did not avoid accusations that some men managed to escape their full share of the more difficult work. No one had grounds for complaint against men who by working hard succeeded in gaining more than others of the work in fields where the crop was light and easy to deal with, provided they did not waste time on other occasions in order to avoid taking their fair share of the work in more difficult crops. It was often alleged that some workers adopted tactics which were unfair. These caused quarrels, if not in the field then at the local when tongues had been liberally lubricated with beer. Disputes were most often between local regular workers and casual men, including Irish workers. There was at all times some ill feeling between regular and casual workers who took the big money jobs away from the former.

In most villages a few men spent the greater part of their working lives on a variety of farm tasks demanding special skills. They moved from farm to farm never spending more than a few days or weeks on any one. Some regular workers living in tied houses and having these skills objected to outsiders doing the work at piece-work rates when they had to work at day rates. The best type of farmer endeavoured to give regular day labourers every opportunity to earn extra money by taking work by the piece. But some tasks had to be given to men not regularly employed on the farm. At busy times of the year the extra labour needed could not be obtained except by letting casuals have the work at piece rates. Routine work which could not be done by the piece had to be done by regular men at day wages. Those who objected to casual workers taking a high proportion of the busy seasonal tasks at piece rates had a remedy, they could move out of the tied cottage and have the freedom and the risks of casual work.

During the Second World War I did a survey of farm labour problems in the market garden areas of the Eastern Counties. I found one farmer in Bedfordshire who had to allow his pigman, in the intervals between morning and afternoon duties with the herd of pigs, to take other seasonal tasks on his or other farms at piece-work rates. Needless to say the general management of the herd suffered, but because of the general shortage of workers the farmer failed to find a pigman who would give his full time to the herd.

As a lad I often heard, at harvest time, complaints from regular workers that badly tied sheaves were the work of casual workers. It was a fairly common opinion that the quality of work done by casuals was inferior to that of regular men. The former at times replied that their work matched the piece-work rate of pay. Much of this criticism was unjustified and due to the antagonism between regular and casual workers. The latter were often not there to challenge those who made the charges. It was true that regular workers who lived in tied cottages had to be more careful about the quality of their work and perhaps employers tended to be more critical of their work than of that done by the seasonal workers. The latter had to be handled more carefully, especially during the busy harvest season, since complaints might cause them to leave when the work was at a critical stage. Casual workers for their part maintained that regular men seemed afraid to stand firm when negotiating piece work rates. They insisted that some farmers used the authority they had over regular men to secure low rates of pay. Regular workers, however, reminded their critics that no one was obliged to accept terms they disliked, least of all men free to take their labour elsewhere without restriction.

Irish labour employed for the harvest received free accommodation, fuel, and potatoes; perquisites not given to regular day labourers or to local casuals. Potatoes occupied an important place in the diet of most farm workers, especially in that of Irish labourers who took advantage of the free supplies to reduce their consumption of other foods. The accommodation provided for Irish workers was not the kind acceptable to local workers; at best it consisted of a specially constructed building while at worst the men had to eat and sleep in the barn or some other outbuilding. Minimum provision was made in the way of furniture, perhaps a table and a form for the men to sit on. Even in 'Paddy houses' the men slept on sacks of straw or chaff with stack sheets or other sacking for covering. They slept in their clothes, and this, with the lack of adequate facilities for washing themselves and their clothes, resulted in their beds becoming lousy before they vacated them at the end of the harvest. Any effort they might make to keep themselves free of lice was bound to fail. At the end of the harvest the task of removing bedding from Paddy houses and barns had to be done by regular workers. Some protested at having to do the work without protective clothing or extra pay. For some local workers this unpleasant task was a good reason for disliking Irish labour. I never heard of any criticism of farmers for failing to ensure that extra harvest labour was properly accommodated.

When carrying crops to the farmstead men worked in teams of six or seven depending on the size of the ricks being built. I have mentioned earlier that the number of boys leading horses between field and farmstead depended on the distance to be travelled. When five vehicles had to be used in order to keep men in the field fully occupied, three, each with a boy to lead the horse, would be moving between field and farmstead. Four or five men in the team normally worked at a piece work rate per man per acre of land cleared. Two piece-workers loaded vehicles in the field and two or three worked at the rick where one passed sheaves from vehicles to another on the rick. He in turn passed the sheaves to a third piece-worker if the size of the rick demanded that five men had to be used. The other two men were the stacker (rick builder) and the binder who laid the second or binding course of sheaves behind these laid by the stacker. Working foremen were usually hired to do the stacking and the head horsemen did the binding. They received no extra pay for working alongside piece-workers nor for working the longer day during the hay and grain harvests. Their normal weekly or yearly wage together with harvest beer and any other perquisites they might be given covered the extra hours they had to work at any time of the year.

On some of the very large farms all the men employed at the rick might be paid at piece-work rates. The practice of fixing the rate on a per man per acre basis allowed any necessary changes to be made in the number of men paid by the piece. Since, however, most of the horses would be employed carting the crops, horsemen were free to help with work at the rick. For this reason men who could build ricks were hired as horsemen, the head man built the rick and the 'second' horseman did the binding. Farmers preferred yearly hired men to build the ricks. They would not be motivated by an urge to speed up the work in order to increase their earnings and could be relied on to pay attention to building ricks which stood firmly and kept the grain dry.

Piece-workers arranged between themselves how their teams should be organised. When, as was commonly the case, only four men formed the piecework team the two pairs did alternate days in the field and at the rick. This could lead to disagreements if one of the pairs considered it was getting too much of the work of unloading sheaves when ricks were at the stage where sheaves had to be thrown up rather than down. Everyone looked forward to a good quick harvest. Before the introduction of the combine harvester country folk judged the prospects for winter work by the number and size of corn ricks in farmyards. For regular workers employed at piece-work a speedy completion of a good harvest meant not just satisfactory weekly earnings during the harvest but steady work throughout the winter. It meant plenty of threshing, of hedging and ditching, and lots of cattle needing food and attention during the winter. With heavy crops harvested in good condition farmers hoped market returns would allow them to spend money on improvements, which gave plenty of work during the winter. Local casual workers also saw in a good harvest prospects for winter work such as tile draining, cleaning out ditches, and hedge plashing. For Irishmen a quick harvest meant a better chance of moving into a potato growing area where female labour was scarce, there they hoped to get more piece-work lifting potatoes. For some Irishmen a quick harvest in England allowed them to go home in time to help with the gathering of their own crops. A quick harvest, however, might be due to a poor growing season and light crops, then, though the quality and price of grain might be good returns could be poor and oblige farmers to cut back on the number of men employed during the winter.

Conditions for harvesting in 1912 were in marked contrast to those of the previous year. Heavy storms flooded large areas in our part of the county. Many fields of grain had been cut before the storms came and the crops were carried away by the flood water. Men had to carry stooks to higher ground. Grain was knocked out of standing crops and the straw was so badly beaten down that scythes instead of reapers had to be used. In fields where crops had not suffered unduly farmers decided to cut them with the scythe; this seemed preferable to waiting until the water had subsided and the land had dried out sufficiently to carry machines. Pigs and other livestock were turned on to the fields to forage among rotting straw and grain. Crops harvested had to be sold for livestock feed if not retained for livestock on the farm itself. Such disasters had serious consequences for farm workers as well as for farmers. Yearly hired men, like father, had their work and wages guaranteed but day labourers, especially casual workers, not only suffered the loss of earnings during harvest but also had to face the risk of little work during the winter months.

Other work done by regular workers at piece-work rates included 'muck carting', that is removing dung from cattle yards. This was done at a rate per score loads or at an overall rate for each stock-yard cleared. The price per score loads before 1914 varied between 2s.6d. and 3s.0d. When this method was used someone, not a member of the piece rate team, was made responsible for keeping a record of the number and size of loads leaving the yards each day. Most farmers preferred to let the work by the second method as it avoided keeping a daily tally of loads removed, it also avoided complaints that some carts were not fully loaded. In a normal working day for horses, that is from 6.30 am. to 2.30 pm. with half an hour break for lunch, two men could fill forty carts, or perhaps one or two more depending on the size of the yard to be cleared, the class of stock wintered in the yard and on the kind and quality of straw used for the bedding.

   In contracting to do the work at a price per stock yard one had to be a good judge of the quality of the straw and dung to be removed as well as the conditions under which cattle had been treated during the winter. By ascertaining depth and surface area men quickly came to a decision on the price they required to do the work. Piece-workers loaded the carts, the unloading at dunghills or in fields was done by horsemen at their ordinary wages.

Chapter 6.

YOUR KING AND COUNTRY NEED YOU
— ON THE LAND

WARS AND THREATS OF WAR have been important incentives behind many of the technical changes in agriculture. In the First World War farmers and farm workers had to accept changes which increased the area under arable crops and also involved changes in methods of production. These, though not always welcomed, could not be resisted. By early 1916 it was clear that action had to be taken by the Government to increase home grown supplies of essential foods. This meant more work for men and horses at a time when fewer of both were available.

The immediate loss of farmworkers after the war started was not serious. Few had previously joined the Territorial Army and only a small number were reservists. Yearly engagements and the distance of most farms from local T.A. Headquarters made it impossible for hired men, especially horsemen, to fulfil the requirements demanded from those who enlisted in the T.A. Recruiting Sergeants might pick up a few recruits for the regular army while doing the rounds of public houses at hiring fairs when men with minds fuddled by beer or those who had failed to get hired decided 'to do seven years in uniform'. But at other times of the year single men hired for the year were not free to take the 'Sovereign's Shilling'. Between the outbreak of war in 1914 and the passing of the Military Service Act in 1916 few men who wished to volunteer got any encouragement from their employers and horsemen could enlist only if they were allowed to break their contracts.

Understandably few farmers living on isolated farms, had any wish to lose men at a time when more land was under arable crops. After 1916 the loss of workers increased the difficulty of finding others unlikely to be called up for military service. It soon became clear that the dwindling labour force could not cope with the extra work. To meet this extra need, soldiers unfit for active service were directed on to farms. In addition the Womens' National Land Service Corps was formed. Women from towns and cities were recruited into the Corps. In our part of Lincolnshire few women had previously been employed as regular workers in agriculture. A few wives from farm cottages and villages gave assistance for short periods at busy times of the year but none had continuous employment throughout the year.

Not all men of military age were directed into the armed services, it was possible, for a variety of reasons, to get exemption from military service. The work of local tribunals, appointed to deal with applications for exemptions, came in for a good deal of criticism, much of it based on false rumours of favouritism. It was assumed by many workers that applications for exemption would only succeed if supported by their employer. It was not sufficiently well understood that men could gain exemptions on personal grounds which had nothing to do with their employment. It was also erroneously assumed by many that exemptions restricted their freedom to move from one farm to another. For single men accustomed to making fairly regular yearly changes in their place of employment, it was important to know that a change of employer did not necessarily affect an exemption. Farmers often gave their men the impression that those who left would lose their exemptions. This was a risk but not a certainty. It depended on whether by making a change a man found himself in work not sufficiently important to justify the continuation of an exemption. Some applicants who failed to get exemptions complained that the Tribunals showed favouritism to sons of farmers. Often these complaints took no account of the positions of responsibility occupied by these sons. One type of complaint, however, asserted that some sons who, before the war, had done no manual work on farms, took key positions, previously occupied by hired workers, in order to avoid military service. In doing so they robbed some labourers of the chance to take work which would have enabled them to get exemption.

The local press, in reporting decisions of the Tribunals, was, by the manner of its reporting, often less than fair to those who gained exemptions. An example of this related to a man named Julius Caesar, who worked on our farm. One local newspaper reported his exemption in the following doggerel:

> 56 BC. Julius Caesar's Roman host
> Landed on the British coast.
> 1916 A.D. Julius Caesar it is written
> Isn't going to fight for Britain.

Whatever interpretation others might put on this announcement the man concerned was upset because in his view it implied that he was shirking his duty. He did not hold a key position on the farm, but genuine personal considerations had persuaded the tribunal to grant an exemption.

Regular farm staffs could, before 1914, cope with the routine work outside the harvesting seasons. When, however, labour became scarce, particularly after 1915, farmers experienced considerable difficulty, even with the help of Irish and soldier-labour, in accomplishing the additional work caused by the larger acreage under crops. As a consequence an increasing number of local women helped and regular workers had to work overtime. This was acceptable to day labourers who were paid for the extra hours worked. Their normal wages had not kept pace with the rising prices of food and clothes, they and their families needed the extra money. The position of yearly hired men was less satisfactory. Father found himself working in the evenings alongside the day labourers for no extra pay. The old bogey of customary hours put him in a different position from the others.

The introduction of a wide range of barn machinery had, before I left school, reduced the amount of manual labour spent by men in charge of livestock. Hand operated chaff cutters had disappeared except on small family farms. At first the driving power for these and other barn machinery was provided by horses and later by oil and petrol engines. A further saving in work for horsemen came when farmers used large mobile chaff cutters. Chaff houses, previously used to store oat straw, now provided accommodation for chopped oat straw in sufficient quantities to last a stable of horses for several weeks. The weekly or twice weekly task of chaffing straw had gone from the work which horsemen had to do after a day's work in the fields. The engines and other barn machinery had also greatly reduced the time and energy spent attending cattle during the winter months when cattle were housed.

During the two years or so before the introduction of statutory regulations of farm wages in 1918 my parents had an extremely difficult time. For men on yearly contracts adjustments in their wages during the contract year, to meet increases in prices of food, clothing and fuel, depended on whether their employer appreciated the unfairness of fixed wages in times of rapidly rising prices. At the yearly hirings provision was often not made in agreements for wages to be adjusted in response to changes in the cost of living. At first people assumed the war would not last many months, this tended to discourage any inclination workers might have to insist on a more flexible method of fixing wages. In any case there was no reliable information available to men living on outlying farms by which changes in the cost of living of rural workers could be fairly assessed. The official index of the cost of living related more specifically to people living in large cities and industrial areas, where consumers were wholly dependent on shops for all their food. In rural areas the situation was quite different. Single men, hired for the year, 'lived in' either with their employer or with a member of the farm staff and in consequence suffered no financial disadvantage from increases in prices of food and fuel. Married men, hired for the year, had some slight protection from rising prices of food if their employer provided a free pig, milk and potatoes. Others with large gardens and potato ground, on farms where they worked, did not suffer the full consequences of the scarcity price of food.

Despite having a free pig and other perquisites my parents found it increasingly difficult to meet household expenses out of the money going into our home each week. My eldest brother was, by 1915, in full time employment on the farm and my eldest sister went into domestic service in May 1915. She had to be fitted out with a specified number of dresses, aprons, caps, boots and shoes of the kind determined by her employer. The cost of the outfit was more than her earnings in her first year in service.

At the yearly hirings for married men farmers tended to argue that the worst was over, that an extra shilling or eighteen pence to the weekly wage would take care of any increases likely to occur in household expenses during the coming year. My parents' position was particularly difficult since they had to provide, under father's contract of service, food and lodgings for three single men on yearly engagements. Before the war mother was paid 9s. per week to cover their cost of food and lodgings. The rate could not be adjusted except at the end of the contract year. In 1916 it was raised to 10s. and other small adjustments were made as prices and wages went up. But at no time during the war years did it fully cover the costs. Later when they had to provide food and lodgings for soldier labour the family resented seeing so much of the home cured bacon being consumed by lodgers. Although we had little difficulty in buying all the bacon we needed the price was not covered by the payments made for the hired workers who lodged with us. I cannot remember whether the farmer or the military authorities paid mother for the board and lodgings of the soldiers but I think the amount was more than that paid for the other lodgers.

Since father was of military age he was reluctant to refuse to take in soldier labour in case he lost his employment and his exemption from military service. We did not have the accommodation to give the extra lodgers the comfort they had a right to expect. It was, however, a situation in which my parents and soldiers had little choice as other lodgings were not available within reasonable distance from the farm. Eventually he was given a second fat pig as part of his wages. Even so with five lodgers and our large family mother found herself depending more and more on the tradesman's cart for bacon. In these circumstances we often found it difficult to hide our dislike of having to share, to an increasing extent, our home cured bacon with men whom father would not ordinarily have had to take in as lodgers. Most farm workers had only the minimal dependence on bought bacon. On many, if not most farms, cottagers had sties in which to rear and fatten pigs for their own consumption. A cottager who could not afford a pig for bacon was pitied by his fellow workers. Some reared and fattened one, others two, depending on the number and age of children at home. A large family with working lads at home needed and could afford a couple of bacon pigs each year. Not all farmers allowed men in tied cottages to keep pigs or poultry. They claimed it encouraged labourers to steal grain, meal and straw. Workers considered the prohibition unreasonable and complained that without home cured bacon they could not live on a labourer's wage. A farm worker in Lindsey, Lincolnshire, told one of the investigators of the Committee on Wages and Conditions of Employment in Agriculture in 1919, that 'it was only pigs and allotments that made the bringing up of nine children possible.' Since the restriction could not be imposed on labourers who lived in free houses with sties those in tied houses took understandable exception to what they considered an implied charge that they, but not other day labourers, might steal grain. Men avoided, if they could, seeking employment and farm cottages from farmers who imposed these restrictions. And farmers who wished to get top class workers had either to allow their men to keep pigs or to sell them fat pigs from their own herds at acceptable prices.

Some farmers allowed men, living in farm cottages and keeping pigs, to have straw from the farm in exchange for the 'muck' which the pigs produced. Others who allowed cottagers to keep pigs provided the men with an acre or half an acre or land, at a nominal rent, on which to produce grain and potatoes, crops which formed a large part of the fattening ration for pigs. In most villages there was land rented out as allotments. Householders could take these half acre or acre plots to produce crops for pigs rather than for kitchens. Only those living in villages and having very small gardens had need to use allotments to produce green vegetables for themselves.

Cottagers who reared and fattened pigs commonly purchased them as weaners in early spring. Until after harvest they fed them on kitchen waste and small quantities of miller's offals. Expenditure on meal was minimal. After harvest the ration was changed to barley meal and boiled potatoes, the quantities being steadily increased until by the beginning of December the pigs were very fat, almost too fat to walk to their feeding troughs. Most cottagers' pigs were killed two to three weeks before Christmas. Everybody wanted 'pig cheer' for Christmas, especially sausages and pork pies. Every village had its pig killer, usually a man with a number of farm skills. At any time of the year he would be found working on seasonal tasks most of which could be done 'by the piece'. Before it became obligatory to observe humane methods of slaughtering animals, pigs were killed on farms and at labourers' cottages. It was never a pleasant sight to see a pig being bled. After killing and dressing the pig was left hanging until cold. It was then cut up and prepared for the salting tub.

Most cottagers preferred to cut up their own pigs, each had his own idea about which parts should be salted down, which used as roasting joints, or for making into sausages and other kinds of prepared foods. Families were torn between two desires, one to have roasting joints and pork pies and the other to have plenty of bacon and ham for the rest of the year. Cutting up the pig was usually done by cottagers in their small living rooms in the evening. The task could be difficult if knives and saws had not been sharpened in readiness. One member of the family held the carcass steady on the cratch* and another held a candle to enable father to see what he was doing. I never liked holding the candle in a dimly lit room. I found that I was either standing or holding the candle in the wrong place. Father constantly complained of being unable to see where his knife was going. If, as sometimes happened, he started cutting up the pig before it was cold and the

---

*Cratch in this context refers to a rack with legs and handles used when slaughtering pigs and sheep.

flesh firm it was very difficult to do the work properly.

First a narrow piece – six to seven inches wide at the neck and four to five inches wide at the tail and tapering to the width of the backbone – was removed from the length of the back of the carcass. This piece was then cut into lengths (called chines) and put into the salting tub for curing. Most cottagers made a point of keeping the neck chine for stuffing in May when sons and daughters, on yearly engagements, came home for their week's holiday. This chine, which had been hanging from Christmas until May, was prepared for the table by cutting into the flesh down to the bone at half-inch intervals, the cuts running from just inside the rind to the backbone. Chopped parsley and perhaps other herbs were stuffed into these cuts; the chine was then wrapped in a cloth and boiled in the copper. When cold, slices were eaten with vinegar and mustard. Most dining rooms in towns in our part of the county had it on offer during May hiring week. There it was taken by those who liked it but had been unlucky in not having it at home. It can still be bought in some of our butchers' shops but with the disappearance of the cottage pig it no longer occupies the place in the life of farm people that it did before 1940.

After the chines, Father removed the ribs and shoulder blades with surrounding muscle which were used by us as roasting joints; the two hams were then cut off and these were put in the salting tub. Next the back muscles – the lean meat of short back rashers – were removed, and were put through the mincer to be made into sausages and pork pies. Other pieces of coarser lean meat were minced and used in the making of haslets. By the time the sides of the pig had been prepared for salting down there was little lean meat left on them. Quite a large part of a pig was used as roasting joints or for a variety of cooked meat specialities which we liked. Neighbours exchanged 'pig cheer' and they all hoped the pig killing season would be spread over several weeks and extend the pleasure of having pig's fry, sausages, and pork pies. Families who reared two pigs each year killed the second one at the end of February or the beginning of March.

Bacon was the most important meat consumed on farms; joints

of fresh meat, if purchased, being restricted to week-ends. Butchers' carts made only one weekly visit to farm cottages and distances between farm and village made it impossible for all but a few to do any mid-week shopping. Our week-end joint usually lasted until Tuesday, the last of it being consumed as stew — what we called forty-to-one because of the quantity of potatoes added. For the rest of the week we had boiled bacon either hot or cold. Meat was consumed at each of the three daily meals in our home, and in most other farm cottages. Except for Sundays we had cold meat at breakfast. Sunday breakfast was a special occasion when we had fried ham, sausages or eggs. Since there was, for most of the year, little variation in the kind of food consumed in our home we looked forward to Christmas time and the fresh pork joints.

Meals taken out at work usually consisted of a piece of meat on a bread crust together with a piece of cake or a jam pasty. We preferred the bread crusts since it was easier to cut our meat on them. The slice of meat was chunky, not thin like that put in sandwiches. We held the meat down on the bread with our thumbs and, since these were often dirty, we used a corner of the crust as a thumb piece. There were no facilities on farms for washing hands other than farmyard pumps and water in ditches. During summer time there was little or no water in the ditches and in the winter the water was frozen over or too cold to encourage us to wash our hands. In any case we had no facilities for drying our hands and thus protect them from becoming chapped.

Chapter 7.

THE LONELY WORLD OF 'COÄD AND SQUADD'*

AFTER SEVEN YEARS AT SCHOOL my parents decided, and the Local Education Authority agreed, that I had received all the education necessary for a potential farm labourer. At that time it was possible for boys to leave school after their twelfth birthday provided they had reached what was considered a satisfactory standard of education, which meant passing simple tests in reading, writing, and arithmetic. When I took the tests, under the supervision of one of H.M. Inspectors of Education, I did a few simple problems in arithmetic, read a passage from a daily newspaper and wrote a letter ordering a ton of coal from a local coal merchant. I was one of a group of about thirty boys from perhaps a dozen parishes in our district, who, in February 1916, took the tests. Three members of out family took the tests between 1913 and 1917 and in consequence left school before the normal school leaving age. I found reading from a daily paper most difficult. It was a severe test for children of parents who did not take these papers. The print was poor and the language unfamiliar to children, certainly to those of farm workers. Writing a letter to a coal merchant gave them little scope to use their imagination and the language of story books. Perhaps some did and I have no doubt the letters gave the Inspector some amusement, assuming he read them. Not all who took the tests passed; but few failed and one suspects everyone was at that time anxious to get as many sons of farm workers as possible away from schools and working on the land.

I was bored with school and very pleased to leave. Having reached a position where teachers appeared to have nothing to offer that was interesting and challenging, I wanted to be away, to

---

* This is Lincolnshire dialect for 'Cold and Mud'

be more active. No-one seemed sufficiently interested to encourage me to go to grammar school; my parents had no wish to treat me differently than my elder brother who intellectually was, at my age, a better scholar and more deserving of a grammar school education. Our parents needed the money we could earn. For my part I cannot remember having any desire to go to a grammar school. We lived a good six miles from the nearest one which could not be reached except by cycle. I did not have a cycle, and in any case the prospect of cycling six miles to school in the winter would not have been to my liking.

Most farm workers considered that education beyond the elementary stage was not for the likes of their children. Few sought for their children, the opportunities they had not enjoyed. Social status, not educational capacity, was the decisive determinant and their sons were born for the farm. A few might escape to the army, if very lucky to the police force, but for most their future was in agriculture. There was no encouragement, in the way of local authority grants, to persuade parents to send children to grammar schools. A few scholarships were provided by local charity organisations but the financial provision was too small to interest parents of large families. In our district only the children of farmers and local tradesmen gained these scholarships.

Country children became so much a part of the agricultural environment in my school days that I doubt if any contemplated spending their working days in any occupation other than farming. Communication between farm and town was so restricted that few children of farm workers had opportunities for meeting those from towns and learning from them all about the kind of work available in shops, offices and factories. For most farm workers the great advantage of placing sons in agriculture was that they would earn more there than in any employment which involved serving an apprenticeship. When family needs compelled parents to place their first son in agriculture it was difficult not to place others in the same occupation, they could not treat one differently from the rest. Entrance to many occupations was by serving an apprenticeship which might last up to five years. For some a premium had to be paid and for most others there was little or no payment in the first year. Employment in towns for most lads from farms was possible only if their parents could afford to place them in lodgings near their place of work. My weekly wage when I first started work was 4s 6d.

Having started work in the middle of the war, with all its shortages of labour, I was pushed into many tasks for which I often had hardly the strength. One had to take risks which at ordinary times would have been avoided; not that farmers worried unduly about the hurt a lad might suffer from overtaxing his strength. It has to be admitted that older farm workers also showed little concern when boys had to do work for which they had neither the skills nor strength. Indeed they often provoked lads to attempt seemingly impossible tasks, fortunately few injuries resulted.

February was not a good month of the year for a lad of twelve to make the change from school to farm work. At any time the change from the companionship of other pupils to the loneliness of farm work was a considerable strain on the equanimity of youths. To this was added in my case the cold, the storms and the mud. I got no comfort from being told by my father that when he left school, at the age of eight, the winters were much more severe. To be out in open fields inadequately protected from wintry weather, with cold hands and feet, with no companions was an experience I had not contemplated in the warmth and company of school and fellow pupils. Of course I knew all about winter on farms but to experience it, unable to escape from it, was something I had not fully appreciated. When working about the farmstead one had some protection from the winter storms, if not from the mud. When out in the fields, however, doing such work as 'turniping sheep' — lifting and slicing turnips, there was no shelter, no protection from the mud. Icy winds did their best to wet dirty hands. When the weather was at its worst one was glad to be employed on work which one could leave and go home until the rain or snow had stopped. But when turniping sheep one had to stay, turnips had to be lifted, cleaned and sliced; the sheep had to be fed. Folds, turnip slicers and troughs had to be moved each day to fresh ground.

At first I assisted the cattleman. This involved carrying straw, hay, roots and concentrates to cattle in the different yards and stalls. I also had to wash and boil large quantities of potatoes for the pigs. Water had to be pumped from wells into troughs serving the yards and stalls. Other work includud cleaning mangolds and swedes stored in clamps in the fields. If the clamps were near a hedge there might be some shelter from cold winds and storms but usually I found this particular task cold and unpleasant. As 'the boy' I was at the beck and call of the cattleman, there was never an idle moment.

One of my earliest summer tasks was to drive cattle and sheep to Spalding market, a weekly market held each Tuesday. We lived nine miles from the town. At that time sheep and pigs were penned at the market but cattle had to be herded on the street. I had to drive cattle and often sheep the full distance on market day. If, however, the fat sheep were in prime condition I took them half way on Monday afternoon. After leaving them in a paddock, I cycled home for the night and returned on the following morning to complete the journey. It might be thought that these weekly trips provided a welcomed change from normal farm tasks. For me that was not so. I found driving sheep to market particularly difficult especially when I had my cycle. This I had to drop from time to time and run after sheep trying to get into fields or running down side roads. I thoroughly disliked the work with all its anxieties. I don't know how many miles I walked and ran each time I did the trip, it depended on how well the animals behaved. Professional drovers had well trained dogs to help in maintaining control whereas I had to depend on my own legs and voice. Drovers became very angry when my animals got mixed up with theirs, it seemed that it was always my fault. When these difficulties did occur it was best to wait until reaching the market before separating them. The problem was less serious with cattle because of the smaller numbers involved and the greater ease of identification.

When I drove cattle and sheep the full distance on market day I did not have my cycle and I travelled home either by train or with my employer in his pony cart. I was instructed to have the animals in the market by 9.00 a.m. This left me with plenty of time, after penning sheep, to look round the stalls before having to catch a mid-morning train for home. The nearest railway station to our home was three miles away, and I was expected to be home for my mid-day meal and at work on the farm in the afternoon. Before making my first journey I was given full instruction by the farmer who asked if I knew how 'to go on at the station'. I had to confess I did not. As children we made few journeys by train and I had never done so on my own. Like most country children I was afraid of trains having been told that one had to be very quick in getting on and out of them, otherwise one was left behind in the one case and carried beyond one's destination in the other. The farmer told me to stay at the sheep pens until he came when he would take me to the station and put me on the right train. When he arrived at the pens he gave me sixpence and on the way to the station initiated me into the way of spending my travel and subsistence allowance. First he directed me into a shop to buy a bun costing half a penny then into another shop to get a penny bottle of 'pop'. When we arrived at the station he showed me where to put the remaining $4^1/_2$d for my railway ticket.

Eighteen years later I had another out of the ordinary experience realating to travel and subsistence allowance. After graduating in 1934 I joined the staff of the Department of Agricultural Economics at Reading University as a student assistant. My salary was £150 per year reduced by five per cent under an economy cut operating at that time. My first work there was in connection with an agricultural survey of Middlesex. Other members of the Department had been working on the survey for some time and when I joined them they had lodgings in Uxbridge. We paid £1.10s per week for our board and lodgings and could make no claim against this cost of mid-day meals taken away from our lodgings. When a student at Aberyswyth I had learned that the staff of the Agricultural Economics Department there received allowances for meals and lodgings when away from their Department. For this reason I thought it a little odd that at Reading University I received no payment, especially for meals which had to be taken away from lodgings. Presumably it was expected that the landlady would provide us with sandwiches. When some of the farm visits took us some distance from Uxbridge we found it was more convenient to have snacks at a wayside cafe rather that travelling back to our lodgings for lunch. When we returned to Reading for the Christmas holidays I enquired whether the University would consider compensating us for the cost of lunches taken away from lodgings. After discussing this request with the University Bursar the Head of our Department told us we would be allowed reasonable out of pocket expenses. I asked what was meant by reasonable and was told to include in our statements of expenses what the lunches had cost. I then asked if five pence would be considered reasonable - that was the amount we had paid for a steak and kidney pie and a cup of coffee — and was advised to charge sixpence. I had expected the University would grant a payment of 2s. 6d. which would have paid for a decent meal. Perhaps it was feared that if a payment of that order had been granted we would be taking lunches away from our lodgings every day and pay our landlady only for bed, breakfast and evening meal. We had no wish to involve the University in unreasonable expense but considered we ought to be treated as favourably as other members of the professional staffs of Universities or more particularly civil servants since the Ministry of Agriculture provided the finances of the agricultural advisory services attached to Universities.

To return to my experiences when attending livestock markets. In Spalding the only protection people had who walked the pavements of the street where cattle were herded was a chain fence erected each market day. It ran the length of the street but did not stop cattle from fouling the pavement. Many people must have been reluctant to do any shopping in that street on market days. Drovers with the aid of their dogs had little difficulty in keeping their cattle herded together while waiting for graders and auctioneers to complete their work. Others, like myself, without dogs or the experience of handling frightened cattle found the task very worrying. War-time controls over the grading and sale of fatstock resulted in more men moving among the animals. This and the noise of barking dogs, of vehicles, and people passing along the street, increased the nervousness of cattle. They frequently jumped over the chain fence and, in their fear, ran wildly down the street. Shopkeepers had to make sure their doors were closed otherwise frightened beast might rush in. No-one knew just how the day would end. I hoped for an early sale and to be free of my responsibilities. Unfortunatley a sale did not necessarily mean immediate release. One had to stay with the animals until arrangements had been made for their removal from the market. Cattle had to be taken to the railway station to be weighed before sale and if after the sale they left the town by rail I might be required to take them to the railway goods yard. Sometimes it was mid-afternoon or later before I was free. On these occasions my only chance of a meal was if my employer came and took charge of the cattle while I paid a visit to a fish and chip shop. Even then I might be expected to eat my fish and chips while herding the cattle. When taking cattle to market it was impossible to know at the start of the day how I would travel home or at what time of day that would be. If I failed to get the mid-afternoon train I returned with my employer if his wife was not with him. If she was I had to take a later train. Taking cattle to market meant I was given a shilling or 1s. 6d to meet my expenses. There was never any encouragement to indulge in riotous spending. It was a great pleasure, to me, to hand the work over to a younger brother when he left school.

Before the introduction of selective weed killers a large amount of manual labour was spent each year on weeding crops. Both hand hoeing and horse drawn hoes were used. As a lad I had to lead the horse for horse hoeing. It was not a pleasant job being both tiring and uninteresting. It was so easy to daydream as one walked at the horse's head. This often resulted in failure to exercise proper control over the horse. When this happened the man guiding the hoe became angry because my neglect made his work more difficult. It was work where one expected wet feet and legs especially during early mornings before the sun had dried off the heavy dew. Cereal crops could be up to a foot in height before work had been completed. I often finished the day wet and miserable. It was worse for the boy leading the horse than for the man since the hoe knocked a lot of the rain or dew off the crop in front of him. At the horse's head boys often had to walk into water laden crops.

Few workers could afford to equip themselves properly against either the weather or farm tasks which involved handling dirty unpleasant materials. Nor did they dare to organise themselves into a trade union in sufficient numbers to be able to insist that employers provided protective clothing and footwear. At the end of the day I hoped for a chance to ride on the horse on our way back to the farmstead but often that was a privilege claimed by the man.

Tasks usually done by boys during threshing included carrying water for the steam engine driving the thresher, also carrying awms and other waste discarded by the threshing machine. Carrying water was a demanding task for a young boy particularly when, due to lack of attention to repairs, the engine wasted a great deal of water. It had to be carried by yolks and two buckets either from pumps or drains. Both could be some distance from the engine. The large buckets, bought by farmers to be used by men, were when full too heavy for lads to carry with ease. Faulty pumps caused delays in filling buckets. The work, however, was most difficult when water had to be carried from drains. Steps cut in drain banks soon became slippery from water spilt from buckets. During severe frosty weather the spilt water froze and added to the hazardous task of mounting the banks. I was often afraid that the slippery steps would cause me to fall and roll into the drain. A lot of water was lost before I reached the engine and at times it was difficult for me to do the work. The engine driver came to my aid a few times, either by putting ashes on the slippery steps or carrying water.

Carrying waste from the thresher was a dirty job, particularly for crops which had suffered storm damage before or during harvest. Threshing peas and beans was particulary dirty work for every one engaged on the task. At the end of the day clothes, boots and hair were filled with fine dust. It was a treat to get home for a good wash. Because of the number of workers in our home and poor facilities for heating water one had either to use cold water or expect a long wait before one's turn at the sink.
    Men employed on threshing usually received two pints of beer each day or, as an alternative, beer money. On our farm the beer was served in half pints at 9.00 a.m., noon, 3.00 p.m., and at the end of the day's work. Boys received neither beer nor beer money despite the fact that often they had to do the dirtiest tasks. One might have thought that farmers would have provided some kind of liquid refreshment. I suppose that there are still regional differences in the kinds of drink workmen take to work, though with the introduction of the thermos flask there may be less variety than when men depended entirely on cold drinks. Our most common cold drink was tea. We drank it, without milk or sugar, in the home at the mid-day meal and took it out to drink at work and at meals taken away from home. It was our practice, and I suppose of most families in our district, to keep a good supply of cold tea in the pantry. After morning and evening meals the tea pot was filled up a few times in order to stock up tea jugs. The importance of beer as a drink at harvest time was a constant subject of debates between strict teetotallers and the rest. Many of the latter insisted that without beer at harvest time it was not possible to sustain a high rate of work output throughout a long working day which rarely finished before 8.00- 9.00 p.m. Some of these men held this view so strongly that they accepted, with reluctance, the inclusion of a teetotaller in their piece-work teams. Not all beer drinkers could afford to take beer to work every harvest day; those who could not had to be content with cold tea, especially if they distained the harvest drinks favoured by 'abstainers'. These included one made with finely ground oatmeal and sugar, another consisted of ground ginger and brown sugar. Some made a drink from a proprietary product called Mason's

Extract: it was said to have the taste of beer, a claim denied by beer drinkers who would not touch it. They called it 'baby's beer'. Some men bought a barrel of beer for the harvest, one of either nine or eighteen gallons depending on the number of workers in the family. Others made nightly visits to their local public house with their gallon or half gallon stone bottles. Some who made these nightly purchases had difficulty in leaving their bottles securely corked until the following day. It was a common sight to see brewers drays visiting farm houses and cottages at the beginning of the grain harvest. In those days we had several local breweries which have now disappeared. Although father had, as part of his wage, eighteen gallons of beer at harvest time, he rarely took any out to the fields. He took his bottle of cold tea and had his beer in the evenings or a glass at any time when he happened to be passing the house.

Lads progressed from simple farm tasks to heavier work involving horses. My first experience of ploughing was in 1917 when I was thirteen years old. It was in a field of heavy clay soil being spring fallowed in preparation for sowing with swedes and turnips. It had been ploughed in the previous autumn and again in April. In June, when I assisted with the third ploughing the field was in a very rough condition with large hard clods of earth which made it necessary to remove the plough wheels. Without these the control of width and depth of furrow depended on the skill and strength of ploughmen. The task demanded more physical strength and prowess than I possessed. From the start I was in trouble being unable to control either the plough or my team of horses. Contrary to the more general practice in other parts of the country we trained our horses to answer to one rein attached to the horse on the land side of the team. When the team was required to move to the left or make a left turn, one gave a long pull on the line (rein) and at the same time told the team to 'coom ether'. When required to move to the right or make a right turn one gave a few sharp jerks on the line and at the same time told the team to 'gee'. It was important to have teams trained to answer the line and calls and for horses making up teams to have been trained to work together.

Owing to the loss of horses to the military authorities it was, by 1917, difficult to have the usual well trained teams. Mine consisted of a couple of old horses which, except during the busy season of spring and autumn, did only cartage work about the farm. They did not respond readily to my calls. I was unable to exercise proper control over the plough, it was either on its side or buried deeply in the soil. The horses seemed to sense my difficulties and added to them by ignoring my commands to stop. My eldest brother, aged 16, and the head horseman came to my help several times. There was nothing they could do; it was obvious the task was beyond my capacity. Father and the farmer must have known I was too small, and too inexperienced for the work. The fact that I had been directed to take the team indicated the serious shortage of trained farm workers at that time. I was glad when father, seeing my predicament, fetched another man to take charge of the team.

Later that year I had to take the same team ploughing stubble after harvest, this time with wheels on the plough. Under these conditions it was much easier to control the width and depth of the furrow. I was not, however, capable of handling either the team or the plough properly. The plough was much too heavy for me, its hales (handles) were too high for one of my stature making it difficult for me to exercise, effectively, the little strength and skills I did have. When, as often happened in my first few days at ploughing that autumn, the plough fell on its side my team dragged it some distance before I managed to stop them. I kept wishing father would send someone to take charge of the cumbersome plough and the badly behaved team. Instead I made an awful mess of ploughing and suffered endless rebukes from both father and farmer. I must confess that I never did attain the skill of a reasonably efficient ploughman.

Another example of boys having to do the work ordinarily done by men occurred during the winter of 1918-19. Although the war had ended agriculture still lacked a sufficient number of regular workers to get through the seasonal tasks expeditiously. Because of this shortage I was given the job of looking after a flock of sheep on turnips. In normal times the work would have been done by a man and a boy. It involved lifting, cleaning and slicing turnips. Wire-netting had to be moved at regular intervals as the land was cleared. Father or the shepherd assisted with moving the wire-netting, the turnip slicer and sheep troughs. I also had a girl to assist on Saturdays when extra work had to be done in preparation for Sundays. I have already referred to the unpleasant nature of much of the work associated with turniping sheep.

Working in wet frosty weather without proper protection caused my hands to become badly chapped. Each evening I went home cold and miserable. I had now reached the stage of wishing to get away from farm work yet dare not refuse to do any task demanded of me by the farmer. My brothers and I often found ourselves called on to do unpleasant tasks and to work at times outside the normal day. Eventually my eldest brother rebelled. Having had enough of being at the beck and call of both father and the farmer he left to work on a neighbouring farm where he felt himself to be a worker with rights of his own. As the sons of a farm foreman of military age living in a tied house we had, during the war years, little choice in what we would and would not do on the farm. We felt that lads whose parents lived in free houses would have refused to do the work required of us except at wages higher than our own. I decided I ought to be paid a wage more in line with that of adults. Father seemed reluctant to ask for my wage to be increased, so I made up my mind, after some encouragement from some of the men, to approach the farmer. He soon told me to 'get on with your work you lazy mawkin, I'll see your father about that'. I felt sure he would tell father that I was getting above myself and was certain he had no intention of giving me the wage I sought. I expected to be in trouble with father. When I went home in the evening I got the impression that father was not unduly displeased with what I had said to the farmer although he gave me to understand that he would, for a little longer, do the asking. Within a short time I got a rise to 2s. per day, by which time I had made up my mind that it ought to have been 3s. Later I learned that father had, when the farmer raised the matter with him, insisted on my right to ask for a wage appropriate to the kind of work I was doing.

Chapter 8.

SOLDIERS ON THE FARM

IN 1916 THREE, and in the following year two, soldiers were directed on to our farm to help with the work. These men lodged with us. Of the first three one was a commercial traveller from Stoke on Trent, another had worked in a biscuit factory in London and the third, an engineer, also came from London. Of the next two one had, as a civilian, worked on farms in the Isle of Man. Both groups stayed on the farm for some weeks. At other times soldier labour came for short periods to help with seasonal tasks.

Some of these men did not have the physical stamina for the work nor the knowledge of how to handle the simplest farm tools. A few made it quite plain that they had no wish to know how to use tools or to do the work. In the early days the men suffered not only the discomforts of working out in the open in all kinds of weather but also the pain from blistered hands. Their daily rate of work output could not be compared with that of skilled men, it would have been unreasonable to expect that it would.

I have no idea how much farmers had to pay for this labour. I assume the men received their army pay and this may have been one of the reasons why there was, on occasions, some discontent among soldiers directed on to the farms. My recollection is that the soldiers disliked having to work alongside labourers who received much higher rates of pay and who were not under the same strict military discipline as themselves. The soldier from the Isle of Man considered he should have been discharged and allowed to go home where agriculture was also short of workers. From the wider point of view the great advantage of retaining control over these men was that they could be directed to areas where the shortage of workers was most serious. Soldier labour being under military discipline could be moved, at will, to any area to meet seasonal needs. The men, naturally, resented that.

Unlike civilian workers, they could not walk off a farm if disagreements arose between themselves and a farmer. When soldiers had to work overtime they expected to get the same rate of pay as local men. If this was granted by farmers local men complained because they felt the pay should be related to the amount and quality of work done. On the other hand if farmers sought to pay soldiers less than local workers they risked losing the help of these men in the evenings.

Complaints by farmers, of the poor quality of work done by soldiers, though justified, often showed a lack of appreciation of the men's inexperience. On one occasion when our farmer complained on this score one of the soldiers quickly retorted that he was not there from choice. This man suggested that had the farmer been obliged to do his civilian job he would have found himself subjected to criticism on the same grounds. This soldier was an earnest worker though inept when tackling most farm tasks. He resented complaints from the farmer who, despite the shortage of labour, spent most of his time riding round the farm or at markets. Much of what the farmer did with his time was essential but one felt he should have spent some time working alongside the men at the busy seasons.

The relationship between soldier labour and regular workers was, in general, good. Only a small number of local workers disliked soldiers enjoying the safety of farm work while their own sons were on active service. Too often they forgot that many of the soldiers had previously been on active service; and because of war injuries or for some other reason had been directed onto farms. A few local workers also made nasty remarks about their local colleagues exempted from military service. They would have approved the actions of those ladies who indiscriminately handed out white feathers to men of military age not in uniform. The only occasion when the farmer and regular men on our farm became really annoyed with soldiers was one summer when three came to help with the hay harvest. From the start these soldiers had no intention of taking the work seriously. When given two-tined forks and directed to help on the hay rick each stuck his right arm between the tines and tried to move the hay by sticking the handle into it. At first this amused the regular men, but when the men showed no sign of settling down to work the regular workers became extremely angry. Like the farmer they wished to get ahead with the work. It was only after the farmer threatened to report the soldiers to the authorities that work on carting the hay got under way.

These particular men, due to their inexperience, made the work more difficult and arduous than it need have been. Experience is required in both loading and unloading vehicles with loose straw and hay. Farm hands know at a glance how a vehicle has been loaded, how each forkful had been laid and the order in which each should be removed if the work is to be done easily. The soldiers tried to lift, from the vehicles, hay which was not free, either because they were standing on it or on hay that was binding it. They also made the work more arduous for themselves and others by their failure to appreciate that when working as a team on some farm tasks it was important to synchronise work movements in order that no one interrupted, by their actions, the work movements of others. The men tended to expend a great deal of unnecessary energy for a low rate of work output. Everybody was pleased when the farmer decided the work would progress more swiftly and harmoniously without the help of the soldiers.

Few of the soldiers directed on to farms in our district had any real appreciation of the wide difference between their own social conditions at home and those of farm workers. Our conditions were, for many, so different that they could not be expected to fit readily into our home life nor into the varied character and conditions of farm work. Most of them must have found the life dull and the work dirty and dreary. The war caused a mix up of people; it gave those, from town and country, as well as from different social classes within their own urban or rural environment, a clearer picture of the way of life of those living in circumstances dissimilar from their own. Many soldiers directed into agriculture must have found life on farms very different from anything they had imagined. Some, no doubt, came from homes where size of family, incomes, and size of homes may have been inferior to those of many farm workers. But town men, if not their womenfolk, had greater opportunites than we had for getting away from overcrowded smoky rooms in the evenings. This difficulty of getting away from the host family made, for soldiers, life on isolated farms so dreary. Those living with us had to walk two miles to the nearest public house, the only place where they could go for company. There they found the conversation too parochial; it could hardly be otherwise since few if any farm workers read books or a daily newspaper. No provision was made by the farmer for these men to spend their free time in warmth and comfort away from our overcrowded draughty living room. He never invited them into his home. In this respect, at least they had the same treatment as regular workers. The Paddy House, which in ordinary times, was empty for the greater part of each year was, during the war years, continuously occupied by one or more men from Ireland. Had it been empty during the winter months our lodgers and other single men working on the farm could have used it for leisure time activities.

With so many people in our house during winter evenings it was not possible for everyone to get near the fire. Younger members of the family went to bed as soon as possible after their evening meal in order to make room for others. Even so, it was not possible to arrange living room furniture so that everyone sat in comfort. The room soom became full of smoke from the fire, men's pipes and cigarettes. This, together with the smells from wet clothes, added greatly to the discomforts which the soldiers must have experienced. Our bedrooms could not be used as a retreat for those unable to leave the house. With two large beds in those occupied by lodgers there was no space for a chair and table.

In the pre-war years father expected hired men living with us to be in the house by 9.00 p.m. except on very special occasions. At first he saw no reason to vary this regulation for the soldiers. He always went to bed at 9.00 p.m. and the rest of us had either to go then or soon after. If any stayed up to read a book it was not long before he called down and complained about wasting candles or oil. It was considered important that horsemen should be in bed by 9.30 p.m. or earlier since they had to be up by 4.30 a.m. Until the war any lodger not in the house by the time father went to bed ran the risk of having to spend the night in the stable or some other farm building. He had to be more reasonable during the war. Men from urban areas, accustomed, when at home, to staying out until 11.00 p.m. or later would not tolerate what they considered as childish restrictions, especially as we lived so far from a public house. I think father knew his rule could not be enforced. Both he and the farmer doubted whether it would have the support of the army authorities. If father had insisted he would either have risked losing his job and house or the farmer would have risked a withdrawal of the soldier labour.

Spring and summer came as a great relief to men cooped during the winter evenings in draughty, smokey living rooms. Then they could get away from the family and live, for a while at least, independent lives. I always thought the soldiers very tolerant in accommodating themselves as well as they did. They may not have expected to find life on farms quite up to the highly colourful, idyllic notions gained from reading some of the popular novels. Even so it must have been difficult to accommodate themselves to our plain country diet, in particular to the excessively fat home cured bacon. The quality of our food was good but it lacked variety. Having to consume boiled bacon, hot or cold, at every meal from Tuesday to Saturday no doubt tested their willingness to endure what must have been an extremely tedious diet. In some other homes cheese might have been offered as an alternative to bacon. It was something we never had. I don't know why, perhaps father did not like it. I have no recollection of hearing the men complain about the food but often wondered in later years, what they had to say about it when away from the house and family. One soldier received supplies of Horlicks Malted Milk tablets from home, perhaps to supplement an unbalanced diet or to take the place of food he could not eat. The local custom of taking pudding before meat and vegetables must have seemed peculiar to these men. It was a practice which was never varied in our house. Some said pudding was taken first in order to save on meat. This may have been the reason with those experiencing serious difficulty in rearing large families on a labourer's wage. But the custom was general and whatever the original reason may have been it had, by the 1900s, become a habit in no way dictated by family incomes. Most labourers produced large quantities of potatoes and other vegetables in their gardens and on allotments and I imagine these rather than pudding enabled families to reduce the consumption of meat. We used the same plate for both courses, which had the advantage of reducing the number of dishes to be washed after meals. I don't suppose it ever occurred to us that the custom would seem odd to other people. Later I often wondered how city men relished eating cold fat bacon off plates that had just been used for hot jam, treacle, or milk puddings. With so many people to cater for mother was terribly over-worked. Living far from a village as well as from their homes, mother felt obliged to wash and mend the lodger's socks, vests and shirts. Our weekly wash day was a formidable task for any one person. The only help given was by children before and after school. In the day time children under school age constantly needed attention causing delays in accomplishing the housework. In the evenings mother was hampered by the number of people occupying the living room. In the winter evenings it was extremely difficult for her to do anything other than necessary repairs to clothes. It was never suggested that lodgers might assist with some of the household chores, for example by assisting with the washing-up after evening meals. Life would have been much easier for mother if, in the evenings, our lodgers had been allowed to use the 'front room'. It was used only on special occasions, rarely more than three or four times in a year. No one used it in their working clothes and except at weekends, or for some particular occasion, farm men did not change out of their working clothes for the evenings. The only time we had a fire in our front room was at Christmas, as a room it was about as useless to our existence as one's appendix.

When, on one occasion, we had five lodgers, making a total of fifteen adults and children in the house, both children and lodgers slept three in a bed. This must have been a further irritation to men who had never previously lived in such crowded conditions. No provision was made in the bedrooms for lodgers clothes. I cannot now remember whether we had sufficient blankets to make everyone comfortable during cold winter nights. The army authorities did not provide the men with pyjamas. Like ourselves they had to sleep in their vest and shirts. The soldiers introduced us to the practice of sleeping in pants. Until then we never wore pants, our trousers being fully lined and our home made shirts half lined. These soldiers must have missed the regular weekly baths to which they had been accustomed at home and in the army. After a hot sweaty summer's day or a cold wet winter one spent in muddy fields they must have longed for a bath in a hot deep tub of soapy water. Even those who, in civilian life, lived in homes without bathrooms must have had opportunities for regular baths. It was a pleasure which they had to forego during their stay with us.

Until the outbreak of war in 1914 we had had little experience of women working on farms. Our corn, cattle and sheep farms offered little opportunity for female workers. Most daughters of farm workers went into domestic service after leaving school. At first they usually went into a local farm house, some might spend all their working days before marriage in rural domestic service. Others, after gaining experience in local houses, went into town households where yearly wages and weekly periods of free time were more generous. It was only at harvest time that one saw more than the odd female working in the fields. One or two labourers' wives might help at other times of the year on such tasks as singling mangolds and swedes and 'spudding thistles' in the pastures. Daughters of small farmers usually helped in the house and fields but only for a year or so immediately after leaving school; they generally expected to find work as shop girls. After 1916 the shortage of local labour forced farmers everywhere to make greater use of women and girls. Some farmers preferred local women to soldier labour, even for work previously considered beyond their physical strength. They loaded and unloaded farm vehicles at harvest time and took teams for ploughing and other cultivation work. Some, rather daringly, drove tractors. The publicity given to work being done by women encouraged many girls to leave domestic service for agriculture. Girls in domestic service did not receive regular weekly wages, most were paid at the end of their contract. Some could, however, obtain small advances during the year. Requests for money to buy necessary clothing for themselves were readily granted, there was less willingness to give advances to meet other demands, for example to meet requests from parents for some financial help. In these circumstances most girls could give little help to their parents except at the end of their hiring. By working on farms their weekly wages made a much needed contribution to the family incomes, especially of families with a large number of small children.

Farm work not only offered girls higher wages than those received in domestic service but also gave them greater freedom in the evenings and at weekends. By the end of the war few girls with opportunities to work on farms went into domestic service. A few parents did discourage their daughters from working on farms. They considered the health risks were too great. There was also a strong reluctance, on the part of some parents, to allow their daughters to work in gangs which included men and youths. They did not wish their girls to be exposed to the language and crudities of some male workers.

At the peak of the programme for increased food production action had to be taken to supplement the supply of local farm labour. Much of the grass land bought under arable cultivation was planted with potatoes, a crop requiring a lot of manual labour per acre. Local labour might manage to deal with a great deal of the planting and the subsequent cultivation work but ouside assistance had to be obtained to lift the crop. A group of girls from the Women's National Land Service Corps came on our farm to help with this work. These girls wore a uniform consisting of a smock, breeches and a hat not unlike that worn by Boy Scouts at that time. It was a uniform suited to the work and added some glamour to the Corps which no doubt was intended as an attraction to the more adventurous type of city girl. Generally the members came in groups to assist with the seasonal work. When they came to our farm they had to be lodged in the barn. I have no knowledge of the farmer being required to make any provision for beds and food and assume the W.N.L.S.C provided these. Many of the girls came from towns, from middle and upper class homes. Few, if any, had personal experience of the conditions under which farm workers lived and worked. The absence of lavatories on farms must have been, initially, a source of embarrassment for those not familiar with the ways of country folk when working in mixed gangs away from farm buildings. Workers, women as well as men, had to retire to ditches or behind hedges to answer calls of nature. If it was a matter of passing water the men might just turn away from their companions, male or female, and relieve themselves. I never worked with members of the W.N.L.S.C. so have no idea how they coped with these problems. Youths, older than me, hung around the barn in the evenings hoping, I fancy, to satisfy their curiosity as to how the girls managed under the primitive provisions available to them.

Owing to the loss of imported onions this crop became very profitable and production spread to districts outside the market gardening counties. Some farmers in our district decided to sow a few acres. It was a crop requiring a large amount of hand labour. In order to cope with this problem a system of share-cropping was introduced. Farmers provided the land and undertook responsibility for all necessary horse work. Workers participating in the scheme did all the sowing, hand weeding, harvesting and preparing the crop for market. The cost of seed and fertilisers was shared by farmer and sharecropper. The scheme was favoured by farmers as a means of holding on to labourers who might otherwise have felt free to move to other farms. Each man employed as a regular worker on farms where the system operated was given an opportunity of sharing with his employer the costs and returns on one acre of onions. My eldest brother had an acre and other members of the family helped with weeding and harvesting. I don't remember the financial results but my impression is that neither farmer nor worker found it a worthwhile enterprise. At any rate the system on our farm was discontinued after one year.

During the winter evenings of those war years country lads searched the skies for a sight of German Zeppelins as they pased over to bomb inland towns. Although far from large industrial areas we always feared bombs would be dropped on our homes. Adults tended to play on these fears. I remember one example of this which ocurred in the winter of 1916-17. One of Father's duties, in the evenings, was to return to the farmstead after tea to weigh out the grain and oilcake rations for livestock for the following day, to do the round of the cattle yards to make sure that the animals were safe for the night and to meet for a talk with the farmer on the work for the following day. My brother and I went in turn to give a hand at the cake mill and at bagging the rations. One dark evening as Father and I returned home from the buildings he dropped his lantern and ran saying that there was a Zeppelin overhead and that it had stopped to drop a bomb. I was scared and ran home as fast as I could. Later I realised it had been one of his tricks but at the time I did think that the Germans had some way of seeing us in the dark. One summer night in 1917 a string of bombs was dropped on a farm a mile or so from where we lived. The only known casualty was a rabbit. All but the first bomb exploded making what we then considered to be enormous craters. Father got up to make sure that livestock in farm buildings and fields had not suffered injuries due to fright. Walking round the fields he came across our four Irishmen hiding under a hedge. There was great excitement on the day the army came to remove the unexploded bomb. Local people stopped work to watch the soldiers at work causing the police to be busy confining sightseers to the ditches. The incident was a topic of conversation for a long time and it gained a great deal in the telling by the more imaginative narrators.

Chapter 9.

TROUBLE WITH THE FURROWS

THE END OF THE WAR, in 1918, brought no immediate let-up in the need for maximum food production from our farms. Nor was there any noticeable increase in the supply of local labour to cope with the extra work. At the May hirings in 1919 farmers found it more difficult than in earlier years to get horsemen with the required skills and experience. Young men, whose parents lived in free houses, became more choosy about the kind of employment that they would take. There was no longer a desire to obtain key posts on farms as a means of avoiding military service. They preferred jobs as day labourers. Men returning from the armed services had no wish to take yearly engagements with all the restrictions these imposed on their freedom to make a change of employer at short notice. They could participate in local sports and other leisure pursuits and they had greater freedom to secure the higher rates of pay which could be earned by casual workers.

Sons of workers living in tied cottages did not have the same freedom, especially where a farmer considered he had the right of first call on the services of all working members of a family living in a farm cottage. In this situation a refusal by a son to accept an offer of a yearly engagement as a horseman might result in his father losing employment and his cottage. No man would have worried unduly about losing his employment but losing a cottage was a more serious matter. The risk was never as great as some workers feared. Sensible farmers had no desire to force the sons of occupants of tied cottages to work for them against their wish; they appreciated that unwilling workers could be most unsatisfactory. At the May hirings in 1919 our farmer failed to get two horsemen required for one of his two farms. As a consequence, although not then half way through my fifteenth year, I had to take the post of second horseman. A lad slightly older, whose father also lived in a farm cottage, was hired as head horseman. I was not asked if I wished to be hired, the decision was father's. As I have mentioned earlier, men who took engagements at the May hirings received a 'fastening penny' making the contract binding. I did not receive this money and never heard that it was given to father. I don't know whether I could have been held to the contract but perhaps as a minor I had to obey the instruction of my father. There was, I believe, a very good reason why my parents wanted me to be one of the two horsemen. They had, by that time, grown tired of having up to five lodgers. With two local lads in charge of the horses the number was reduced to three.

I still had a great deal to learn about handling horses as well as about managing implements and tackling cultivation work. To make matters worse my dislike of horses meant that I was not a keen pupil of that branch of farmwork. I was completely ignorant of the technicalities of marking out a field for ploughing. The first furrows, the ridges of ploughed fields, were drawn at intervals of roughly twenty-two yards. It was important to draw these furrows parallel to each other in order to avoid wasting time on short turns with the plough. When drawing my first ridge I paced out twenty-two yards along the line of both ends of the field, failing to note that they were not parallel to each other. Consequently the width of land to be ploughed was much wider at one end of the field than at the other. That was my first mistake. My second was to direct my team from one end of the field to the other by reference to one marker flag. I had seen experienced ploughmen draw a straight furrow with only one flag placed at the distant end of a field and mistakenly assumed they had done so by the simple act of directing their eyes on that one flag. The head horseman, only slightly older than myself, was equally ignorant of how to draw a straight first furrow. No-one had thought it necessary to make sure we knew how to mark out a field. I certainly had not appreciated that the only sure way to plough a straight furrow with only one marker flag was by keeping that flag constantly in line with some other distant object. Having failed to do so I finished with a furrow having a nasty curve. These two mistakes caused me to have to make several short turnings with my plough and in consequence to waste a lot of time.

In those days famers and ploughmen frequently disputed whether it was possible to plough an acre of land in an ordinary plough day of $7^1/_2$ hours. It was not often that we managed that rate of work output. With a furrow of only nine inches width one had to keep the team moving sharply all day to cover an acre of ground. Horsemen never liked urging their horses at speeds other than their normal pace which for Shires carrying too much flesh could be very slow. Most ploughmen sang or whistled at their work. With three or more teams in a field and the men singing or whistling it was, I must admit, a most pleasant sound. One of their most popular songs was the **Ploughman's Song**, especially the following two verses:

> Our master he came to us, and this he did say:
> 'What have you been doing, boys, all this long day?
> You've not ploughed an acre, I swear and I vow,
> You're damned lazy fellows that follow the plough.'

> I stepped up to him and made this reply:
> 'We've all ploughed an acre, so you telled a damned lie,
> We've all ploughed an acre, I swear and I vow,
> For we're all jolly fellows that follow the plough.'

Of course men wasted time in all sorts of ways. If neither father nor the boss was about and the day was pleasantly warm we took more time for lunch than the regulation half hour. We also gossiped at furrow ends and larked about at odd times during the day. Then in order to make up for time wasted we increased the width of the furrow. We had then to reduce the depth of the furrow in order not to increase, unduly, the work load for the team. Only small adjustments could be made to the furrow since farmers quickly distinguished between a nine by six inch and a ten by five-and-a-half inch furrow. The latter lay more on its side than the former. Measurements of width and depth of furrows could be done quite easily without interrupting the work but our farmer often stopped the teams to make a check. We knew it was his way of telling us that adjustments had been made to the ploughs contrary to instructions. We watched the comings and goings of the farmer and father and unless caught unawares ploughs would be properly set when they were in the field.

Prior to the introduction of petrol driven tractors in the First World War the larger class of farmers in our county employed steam ploughing tackle to do some of their autumn and spring cultivation work. Few farmers had sufficient arable land to justify owning a set of their own, most employed contractors to do the work. Each set of tackle had two engines with cable drums; two ploughs; a cultivator, called a drag; a water cart; and a living van for men in charge of the outfit. The team mostly consisted of a foreman, two engine drivers, ploughman and a cook. The men worked, weather permitting, all the hours of daylight. Each man in turn took time off for a meal. One of the ploughs with six breasts ploughed to depths of ten inches, the other with three breasts ploughed to deeper depths and had wider furrows, and was used in the silt areas where potato and vegetable crops were produced.

Farmers provided a man and horses to convey water and coals to the engines. I did this work once. It was a task no one liked because one had to be at it all the time and, in the summer time, for a very long day. Water had to be pumped by hand from drains, ditches or ponds. Towards the end of a busy spring or autumn season the pump attached to the water cart was most likely to be defective in which case a lot of time and energy could be spent filling the cart. In the summer one might have to go some distance to find clean water. If this was associated with a defective pump there was always the danger of being unable to keep the two engines fully supplied with water and coal. The men employed by the contractor were paid an acreage bonus in addition to their normal rates of pay. Understandably they became annoyed if work was held up due to lack of water or coal or if, as a consequence of being supplied with dirty water, work had to stop in order to wash out boilers. It was customary to do the latter on Sundays. Carting coals and water could be even more irksome if the horses used for the work were easily frightened; it could cause a great deal of time to be taken up in getting them to place carts near the engines. If, as happened on occasions, engines had to be moved to the water cart the drivers became very annoyed and might insist that another team of horses had to be used for the work.

Horsemen hired for the year disliked the work for another reason. In the summer the contractor's men often started work as early as 4.30 a.m., this meant horsemen had to be up in time to feed and groom their horses and be ready to answer early morning calls for water or coal. In order to avoid the risk of being held up at the start of a day's work drivers liked to leave their engines fully supplied with water and coal at the end of a day. This meant the man doing the carting might not finish his day before 9.00 p.m., and often it would be later, before he had fed and watered his horses and turned them out on the pastures. If this work was done by horsemen on yearly contracts the very long working day came within the definition of customary hours. Some farmers, and perhaps contractors, rewarded these men for the extra time worked. Many, however, had only their normal wage. On farms where this was most likley to happen horsemen were pleased when day labourers did the work, then farmers had to pay overtime for the extra hours worked. It was not easy to get men, other than those living in tied houses, to do the work. We were always pleased to see these iron horses make their smokey way from our farm.

Before the introduction of statutory control of wages most farm workers, other than single men hired for the year, had few paid holidays. I don't remember that we had Christmas day as a paid holiday, but I know we did have the afternoon of Good Friday off without loss of wages. After 1918 we had, I think, two paid public holidays, of which Christmas day was one. My parents could not afford to take holidays other than those granted without loss of pay. Before their marriage both had – as yearly hired workers, father as horseman and mother as a domestic servant – had the week's holiday in May and three or four days at Christmas or the New Year. This I think must have influenced their decision that we children should, when old enough to have our own money, take a holiday each year. In 1919, and again in the following year, two brothers and myself had a week's holiday in May. In 1921 when we moved to the Boston district the system of intensive cropping made it impossible for us to take holidays during the planting and growing season. To have insisted on taking a holiday in the spring or summer seasons would have put our jobs at risk. More important, with the fall in wages after 1921, we could not afford to stay away from work.

Even before 1921 we did not have sufficient spare money to take holidays which involved staying at hotels or boarding houses. We went to stay with relatives, with grandparents or with aunts and uncles. On the occasions when a younger brother and I had holidays we spent them with our maternal grandparents who lived about four miles outside Skegness. The East Coast rarely offers a warm welcome to visitors in May whatever it may do in August. During one of the holidays we went into Skegness but the cold winds soon persuaded us to return to the warmth of our grandparents' home. On our first holiday we travelled the sixteen miles to Boston by carrier's cart. That part of the journey was uninteresting because, sitting in the cart, we saw little of the villages and farms. At Boston we were met by grandmother who had hired a pony cart to take us the twenty miles to her home. We had not previously been so far from home and most of the countryside was new to us. On the second occasion we did the full journey on cycles. This gave us an opportunity to stop and get our fill of the things we found interesting on the journey. Young people today, would consider our holidays very dull affairs but we found them satisfying: they took us away from home and, for me at least, from the boredom of farm work. We considered ourselves lucky since few lads of our age other than those on yearly engagements had holidays.

My first exciting day's outing was in 1920 when three ex-army lorries took a large party from our district to Skegness. A local man had bought a number of these lorries and set up business as a haulage contractor. In addition to moving farm produce and other goods he took parties on outings. Motor coaches had not at that time penetrated into the country areas of Lincolnshire. The lorries had solid tyres and forms* provided the seats. It was a hard, and on bends in the roads, an unstable journey. The pace was a leisurely twelve miles per hour. Even those with an uncontrollable urge to get to the seaside found the speed, at times, a strain on their nerves. Most of us remembered, for some time afterwards, more about the journey to and from Skegness than about how we spent our time taking the sea air. It was my first visit to a seaside resort during the holiday season. Like most of the party we had no idea how to organise ourselves so as to get the maximum amount of pleasure from the minimum amount of moving from place to place. There seemed to be so much we wanted to see and do, we rushed from stall to stall, from one amusement to another. Long before the end of the day I was tired out.

Although we had at that time good wages some of us were reluctant to spend much on amusements. We had heard many stories about the dishonesty of cheap jacks and stall holders; these persuaded me not to make any bids for the "gold watches" or to try my luck at catching tin fishes in tanks of water. The bathing machines had a busy time. Four of us hired one and bathing costumes much the worse for wear. It was exciting to strip down, and put on for the first time a bathing costume for our first dip in the sea. Dip is not the right word for I only walked knee deep in the rolling sea. Mothers with small children must have had a tiring day especially when fathers and older children had no wish to take on some of the responsibility. Older members of the party moved about in groups, few daring to go off on their own for fear of becoming lost in the crowds. A few stops had to be made on the journey to and from the sea. Some men needed liquid refreshments. Mechanical troubles with one of the lorries caused some delay but fortunately none of these halts caused serious loss of time on the outward journey. Only Mums regretted the few stops at public houses on the journey home. They had fractious infants that could not be pacified.

---

*forms - long seats without backs

Chapter 10.

'WORTHY OF HIS HIRE'

THE INTRODUCTION OF statutory regulation of farm wages following the passing of the Corn Production Act, 1917, was the beginning of important changes for farm workers. The immediate consequence was a substantial increase in wages and a reduction in the length of the working week. The Agricultural Wages Board, set up under the Act, introduced greater uniformity in the contractural relationships between farmers and their workers. As a result there was a rapid decline in yearly engagements of both married and single men.

The average weekly rate of pay for ordinary day labourers in Lincolnshire was, in August 1914, 16s.8d. Over the next three years it went up to £1.3s.9d. and just before the first Wages Order for the county became operative, in 1918, had increased to £1.5s.

Thus during the period from 1914 to 1918 farm wages increased by about 50 per cent. In contrast the official statistics for the whole country showed that the cost of living had doubled. The first Wages Order for the county increased the weekly wage of ordinary labourers by 9s. making it, relative to the cost of living, equal to that paid in 1914. Over the next year the minimum wage was increased to £2.0s.6d. and, by 1920, to £2.8s.6d. This was the highest rate fixed for ordinary labourers in Lincolnshire. It was almost three times that paid in 1914 and nearly 43 per cent above the first rate fixed for the country in 1918.

In contrast prices of food and other consumer goods increased, between 1914 and 1920, one and a half times, the greater part of the increase occuring during the first four years of this period. Farmers insisted that workers had been much better off under the old system of individual bargaining. The facts, as the following summary shows, did not support their claim.

Year	Wages	Cost of Living
1914	100	100
1918	151*	
1918	206†	203
1919	248	215
1920	294	249

* Before † After, the introduction of minimum wages

We did not need the facts to be formalised in this way to convince us that the Wages Board had greatly improved our standard of living. For the first time married men with families, with the help of overtime earnings, saved a little money. The Wages Board conferred other benefits. By fixing the length of the working week and giving us the Saturday half-day we had, at last, gained what our Union had been agitating for over a considerable time without success. Under the old system of individual bargaining there had been variations in the wages paid, not only between workers on neighbouring farms but often between men employed on the same farm. One accepted differences between men doing disimilar work and having differing responsibilities but when groups of men working alongside each other had no knowledge of what each was paid there was bound to be a good deal of suspicion of favouritism. There was, for example, the suspicion that the man in the team who always seemed to be the pace setter was being paid a little more than the rest. It was generally thought that the controls rid the industry of some of the variations in wages paid to the same class of worker but one never knew for sure if this was so. Statutory control of wages did not of itself eliminate these variations since a farmer was always free to pay any or all his workers more than the rate fixed by the Wages Board. All one could assume was that each received not less than the legal minimum. One would expect the scarcity of highly skilled workers during the war and immediate post-war years to oblige farmers to pay these men more than the prescribed minima. Men not affected by the provisions of the Military Service Acts, who lived in free cottages, used the shortage of labour to secure better wages and conditions of employment than those provided by statute. Others liable for military service, especially those in service cottages, hesitated to ask for wages commensurate with their worth.

There was considerable opposition to the Wages Board from farmers. Many of their objections showed a real, or pretended, ignorance of the provisions of Wages Orders. Some complained of having to pay incompetent workers the same minimum rates as those paid to highly skilled labourers. If, as seemed to be implied, farmers paid skilled labourers no more than the legal minimum they were fortunate. This complaint also ignored the fact that an employer could apply for an exemption from the normal fixed rates for workers with physical or mental disabilities which reduced their capacity to do the work of ordinary workers.

The first Wages Order for Lincolnshire came into operation during the grain harvest in 1918, consequently we did not start taking the weekly half day until the harvest had been completed. It meant, that for those on day rates, the time worked after 1.00 p.m. on Saturdays had now to be paid for at the higher overtime rates. Previously we had worked the normal hours on Saturdays which meant, weather permitting, harvest work did not stop before 8.00 p.m. There was never a chance to spend Saturday evenings at the cinema during harvest. In 1918 we stopped work at 4.00 p.m. on Saturdays. Men working at piece-work rates had no particular wish to break with tradition. They wanted the extra earnings and to get ahead with the work. Irishmen in particular wished to get harvest work finished in order to get home and harvest their own crops or to move into another area and assist with lifting potatoes. Men employed at day rates, especially yearly hired men with their minimum rates fixed to customary hours, had no wish to work the same hours on Saturdays as on other days of the week.

Some farmers insisted that the Wages Board was unnecessary as they already paid more than the rates fixed. They may have done, but as indicated earlier the average weekly wage paid before statutory controls came into operation showed that in general workers had failed to get, by individual bargaining, wages commensurate with the rising cost of living. In their public criticism farmers chose to ignore that the Wages Board prescribed only minimum rates of wages leaving farmers and workers free to bargain for better conditions if they so wished. Another criticism was that the Wages Board damaged the good relationship existing between master and servant. Apart from the fact that it gave farmers something else to grumble about I saw nothing which indicated a worsening of relationships after the Wages Board became operative. Farmers resented the growing strength of trade unionism among farm workers and the increased authority which the unions gained over workers following the introduction of the minimum wage. Contrary to the views expressed by some farmers an increasing number of workers accepted that their interests would be better served by a strong trade union and a Wages Board. They appreciated the necessity for an organization which enabled them to select their representatives to serve on the Wages Board and on District Wages Committees. In 1917 the National Agricultural Labourers and Rural Workers Union had just over 15,000 members. Over the next two years membership increased to nearly 127,000. Similar increases occurred in other trade unions serving farm workers. Despite this growth in membership it was not possible to find a sufficient number of workers able and willing to serve on all District Wages Committees. One reason for this was the reluctance of some farmers to allow men time off to serve on these Committees. If the men selected to serve on these Committees came from a farm where several workers were employed, frequent absences of one man for short periods caused little difficulty for the farmer. A small farmer might, however, experience considerable inconvenience if his only day worker had to be allowed time off to attend meetings.
    Complaints by some workers of having been threatened with

dismissal if they accepted invitations to serve on District Wages Committees were sufficiently numerous to oblige the National Farmers' Union to issue a statement asking their members not to place obstacles in the way of workers invited by the Wages Board to serve on these Committees. Threats of dismissal only added marginally to the problem which trade unions catering for farm workers had, in finding bona-fide farm workers willing and capable of representing their colleagues on District Committees. In order to get the required numbers the unions nominated members of other trade unions. Farmers claimed this was a clear indication that farm workers, like themselves, did not want statutory control of wages and conditions of employment. They asserted that the industry had been burdened with regulations no one wanted. They complained that their representatives had to negotiate wages and conditions of employment for their workers with persons lacking any appreciation of the close personal relationship which existed between themselves and their employees. In their view bringing into the discussions the ideas of men who had spent the greater part of their working lives in manufacturing or servicing industries could do great damage to the good relationship which one had in agriculture where many of the employers worked with their men.

When workers complained of some particular action taken by the Wages Board farmers told them that so long as they had as their representatives on District Wages Committees, workers from other industries they could expect to be troubled with unsatisfactory regulations. From the frequent remarks made by farmers one might have thought that none of the workers' representatives, serving on the Wages Board and District Committees, had ever worked on farms; that those selected from other industries and occupations had no knowledge of the needs and aspirations of farmworkers. The number of workers' representatives who were not bona-fide farm workers was a very small proportion of the total serving on any one of the District Committees. In areas where farmworkers' trade unions were strong, officials from the unions were the only persons not directly working on farms who served on these Committees.

The contractual relationship between farmers and their workers was, prior to the 1917 Act, in a large measure informal. In general this had served the best interests of both sides. The small farmer, employing one or at most two regular workers, spent a large part of each day working alongside his employees and in the evenings might be seen having his pint of beer with them. Action taken by the Wages Board in no way interferred with this good relationship but it did recognise that, as was often the case on this class of farm, the level of wages tended to be below the general average.

The Board fixed both weekly wages and hours for ordinary labourers but only the wages of horsemen, stockmen, and shepherds. For each of these three classes the minimum rate of wages was related to their customary hours of work. This was not entirely satisfactory. Many workers in these categories considered it left too much to be decided by farmers rather than jointly by themselves and their employers. At first they felt that the risk of being called up for military service placed them in a weak bargaining position. Fortunately the war ended soon after these controls came into operation and this enabled men to be more firm in their demands.

Some farmers assumed 'customary hours' meant the weekly hours normally worked by special classes on their particular farms before the introduction of controls. This was not accepted by workers who wished to secure greater local uniformity in weekly hours. They maintained the term meant the weekly hours generally worked by each of the classes in the district covered by the particular Wages Committee in the immediate pre-war years. In some instances the customary hours had been increased during the war owing to the scarcity of labour. For example many shepherds and cattlemen no longer had the help of boys at weekends. The old traditional three hours Sunday work became four or more.

Farmers maintained that since District Wages Committees would know, or could ascertain, the weekly hours most commonly worked by each of these special classes these would have been prescribed in the Wages Orders had that been their intention. The fact that no such provision was made implied that customary hours was a matter to be negotiated between each farmer and his men. Workers, for their part, wished to use the authority of Wages Orders to oblige the worst type of farmer to observe the standards set by the best in their district.

It was difficult for District Wages Committees and the Wages Board to give precise instructions on weekly hours for these workers since these varied over the year and at any time of the year they might be affected by some unusual circumstances. Perhaps the most satisfactory way of dealing with this problem would have been to relate the weekly wage to a specified number of hours and at the same time fix an hourly rate of pay for any difference between the prescribed weekly hours and those worked in any week.

Shepherds and men in charge of cattle gained little from the reduction in the length of the working week of ordinary labourers. There was no reduction in the amount of work both had to do on Saturday afternoons and Sundays. During the lambing season shepherds had to be on continuous duty day and night. Any reduction which they gained in the weekly hours worked came as a consequence of the small reduction in the length of the working days from Monday to Friday.

Horsemen on our farms gained a reduction of not more than three hours per week from the introduction of the Saturday half day and nothing from any reduction in the length of the other days of the week. On these days their hours remained as before from 4.30 a.m. to 7.00 p.m. with one and a half hours off for meals. On Saturdays field work stopped in time for horses to be stabled and unharnessed by 1.00 p.m. Work in the stables on Saturday might finish at 4.00 p.m. instead of 7.00 p.m. The men however, had to visit the stables before going to bed to make sure everything was in order and if necessary to put more food in the mangers.

No action was taken by our District Wages Committee to regulate the wages or hours of foremen. Since their status and duties varied from farm to farm any attempt to formalise these into uniform regulations over the whole range of size and types of farms would have been unrealistic. The commonsense view was that since minimum rates had been fixed for other grades of workers it could be left to foremen to secure wages and other conditions commensurate with their varied responsibilities. The status and security of some, however, was too weak for them to gain, by individual bargaining, rates of wages which took fully into account the hours worked beyond those of ordinary labourers. Some foremen complained that the increase in their wages over the period from 1914 to 1921 had been less satisfactory than that of day labourers. They maintained that their pre-war differentials had been eroded because farmers tended to think of these as a fixed amount whereas they considered the pre-war percentage variation between their wages and those of men under their charge ought to be maintained. Some argued that because of the extra responsibilities due to changes in systems and methods of farming, and also the increase in farm machinery, percentage differentials ought to be increased.

My father became very dissatisfied with his position as foreman. He found himself working an increasing number of hours each week for a wage which over the year was below that of ordinary labourers. The latter had, because of the shortage of workers, been able to obtain more work at piece work rates. This had encouraged them to work a longer week and secure a substantial increase in weekly earnings.

Mother was anxious for a change, she wanted to be free of the lodgers. By 1920 five members of our family had left school and the amount of weekly income going into our home encouraged my parents to think a change would be to their advantage. Mother, if not father, wished to live near a village and to make more frequent visits to shops. For nine years she had been obliged to depend to a large extent on tradesmen's carts and the 'carrier'. Living long distances from shops had been a great source of worry and annoyance since it had not been possible to take full advantage of the little extra rationed and unrationed goods that came into the shops. We knew from our talk with workers who came from villages that one had to live near shops and make frequent visits to them if one wished to get a share of goods in short supply.

We lived at the end of the rounds of both baker and grocer tradesmen's carts and in consequence had little choice of the things we needed. By the time the baker reached our house he often had insufficient bread for our needs. Mother had either to make soda bread to help us over to the baker's next visit or one of the family had to make the journey of nearly four miles to the nearest village bakehouse. None of our tradesmen came from this village and in consequence we could never be sure that they would sell us bread and other goods we wished to purchase. Nor could we know whether the longer journey to our own suppliers would prove more satisfactory. No member of the family liked making these trips after a day's work in the fields especially in the winter when cycling could be difficult.

Chapter 11.

No Social Life

WHEN IN APRIL 1921 FATHER DECIDED to move and take work as a general labourer we all welcomed the decision. Those at work were pleased to have a change and hoped it would give us the opportunity to choose work on farms other than that where Father might be working. Younger children hoped the move would take them nearer to a school. The move was to within a mile of the village of Sutterton and within seven miles of Boston. Mother was glad to be near a good public road, nearer to village shops and close to a railway station. Between 1917 and 1921 we had lived in a house some distance from a public road and approached by a badly rutted farm road.

Initially my father had intended to seek employment with Richard Leggot and not C. W. Hardy from whom he did get an offer of work and a house. When he had arrived in Sutterton on his way to contact Mr Leggot's foreman he had met an old acquaintance who, until retirement, had worked for Mr J. T. Ward of Carrington, a large farmer with farms in both Carrington and Holbeach St Marks. He had told Father of another old acquaintance, Harry Clements, who had worked for Mr Ward and was then foreman for Mr C. W. Hardy who had farms in Sutterton and Wigtoft. Father decided to get in touch with him and accepted a job which put him in charge of a gang of female land-workers. A house went with the job. I worked for three years for this farmer but saw very little of him except in his car. The foreman, acting presumably on directions from Mr Hardy, had full control of the day-to-day tasks of the workers. Much of Mr Hardy's time was spent visiting other producers of market garden crops and buying, whenever he could, crops that were ready for the market.

Had mother seen our new house before we moved I am sure she would have been less happy about the change. It had only three bedrooms and the downstairs rooms were too small to accommodate us all in comfort. It was not an old house and had had only one previous tenant. It had been built in the corner of a field, but no-one had bothered to plant a hedge or put a fence round it and the garden, and both the employer and the tenant had neglected the house. The latter and his family had been too busy going out to work to have time to make their home a tidy comfortable place in which to live. No one had troubled to provide accommodation for cycles or farm tools. One of the first things we did was to erect a shed for these.

At that time we had no country bus service and for those members of the family who could not cycle the railway station, which was about a quarter of a mile from our house, was, with the exception of market days, their only means of visiting Boston. Each Saturday evening throughout the year a large number of women came out of the station laden with baskets and parcels. These shoppers not only saved more than the railway fare but also had a wider choice of goods than could be had from village shops. With the decline in farm wages after 1921 it became important for wives to hunt around for bargains. These weekly trips also provided, for women tied to their homes at other times of the week, a welcomed opportunity for some local gossip. For mother the change was most welcome as previously she had been so frustrated by the lack of a chance to do competitive shopping or to have a chat with neighbours and meet relatives.

The move released my brothers and myself from the restrictions and inconveniences of being the foreman's sons who could be called on to do emergency jobs outside the normal working day when other men were not available. For too long we felt we had lived too near our work, too near the foreman and the farmer. The change, however, did not give us all we had hoped for. When seeking this change Father had mentioned that he had sons and daughters who would be looking for work. We would have preferred to find work for ourselves on farms away from that on which Father worked. In 1921 farmers in the intensive arable cropping areas appreciated having their cottages occupied by men having sons and daughters available for farm work. We disliked being a part of Father's bargain, it was a restriction on our freedom. We feared that if at any time we wished to make a change of employer it would cause difficulties for Father. The move did not, as I had hoped, mean less horse work for me. It seemed the only way of escaping from the bargain Father had made was to leave home. That would certainly have meant taking a yearly engagement as horseman on another farm. I preferred to stay at home hoping that I would soon be given other work, perhaps in one of the piece work teams employed on that farm.

The system of farming was in sharp contrast to that practiced on our previous farm. A much higher proportion of the land was under arable crops, in particular cabbage, potatoes, sugar beet and other root crops grown for seed. There was a fair amount of double cropping, that is, producing two crops per year from some of the fields. The arrangement of the working day was also different. Instead of working from 6.30 a.m. to 5.30 p.m. with an hour and a half for meals we now started work half an hour later and finished half an hour earlier with only one half an hour meal break. The opportunity for a hot midday meal was gone, we had to do with a packed lunch at 11.00 a.m. taken at our place of work. Everyone working on the farm, including horsemen, had their meal break at the same time. We appreciated being able to stop work earlier in the afternoon. It, together with living nearer to a village and town, made it possible for us to get away from home in the evenings.

There was a considerable amount of horse work connected with potatoes and vegetables, and with root crops grown on contract for seed. Most of my time from March to October, during the first two years on this farm, was spent with horses on work connected with the cultivation and harvesting of these crops. The normal working day for horses was longer than I had previously experienced. We took the teams out at the start of the working day and did not return to the stable until the end of our day. During the lunch interval the horses were given their nosebags but no water. There was a good deal of row cultivation work during spring and early summer done with teams consisting of one horse and a schuffler or cultivator. I particularly disliked schuffling potatoes. One had to walk behind the implement to control it and the horse. The schuffler was so constructed that unless the soil was very firm and heavy one had to put some lift on the hales (handles) to stop it burying itself in the soil. this imposed a great strain on one's arms, especially for those below average height. Because of the large acreage of potatoes grown on this farm the work could be more or less continuous from mid April to the end of June. It was hard on the feet especially on a hot June day when the heat from the soil played havoc with them. I was always pleased at the end of a day to get home and put my feet in a bucket of cold water.

On large farms a great deal of the manual work on planting, weeding and harvesting was done at piece work rates. It was possible for some regular workers to spend the greater part of each year on piece work. Few labourers living in free houses in the Boston district stayed for long periods with a farmer unless they had plenty of opportunities for piece work and to earn much more than ordinary day wages. To my annoyance I had few chances of earning extra money during the normal working day. I did, however, secure, as did others in my situation, piece work in the evenings. During the early summer months men, women and children with free time could, if they wished, find a farmer needing workers to pull peas. A large acreage of peas for the green pea market was grown in the district. One could usually find a farm not too far from home where a crop was ready for pulling. We went, when free, with buckets or baskets to the farm of our choice, picked up a bag from the man in charge and started work. In the 1920s the rate of pay varied from 10d. to 1s. per pot of 40lb. Pickers came and left at times convenient to themselves. Women and girls formed the greater part of those employed on this work in the daytime. Children came to help after school and after tea men joined in the work. On occasions farmers with poor or weedy crops had difficulty in getting pea pullers unless they offered higher rates of pay. In the height of the season when market prices were at the seasonal minimum few light or dirty crops justified paying the higher piece work rates. Indeed in some years acreages and yields forced market prices down so low that some crops had to be harvested dry. Some farmers paid pullers for each pot as it was weighed and checked by the man in charge. This left the workers free to stop work when they pleased. It suited those who had a reputation for shifting from one farmer to another in order to take advantage of the best crops and rates of pay. Owners of poor crops often found themselves without sufficient pullers to get the crop marketed before the peas became too old. They paid only at the end of each day, making exceptions in the case of women, who, for domestic reasons, could only give short periods to the work.

In the days before the introduction of machines for planting and lifting potatoes this work was done mainly by women and children. In some isolated districts regular and casual male workers did the work at piece work rates. It provided women and school leavers, in the potato growing districts, with work for the greater part of spring and summer. In the silt areas of Lincolnshire planting could take up to five weeks and lifting, commencing with the early varieties in June, continued with few breaks until the last of the maincrop had been lifted, which might not be before the end of October or later.

Women living on farms and in villages in our area could find fairly regular employment on farms from the beginning of March until the end of October. If work under cover was available the period might be longer. The demand for female workers and teenage boys slackened off after potato lifting had finished. Few of these workers found employment during the winter months. This did not bother housewives interested only in earning a little pin money during the spring and summer. Young girls, boys and widows entirely dependent on their earnings suffered considerable hardship each winter due to the long periods of unemployment. It was difficult for many to build up savings to help them through the winter. Before 1936 farm workers had no formal financial protection against unemployment. Those unemployed either depended on their savings, on their relatives, or had to seek relief from their parish.

Girls living at home and working on the land had a more difficult time than their brothers. For in addition to the insecurity of employment and poor earnings, indeed perhaps because of these, they had to help with the household chores at all times of the year. Sons came home in the early evenings expecting their meals to be ready, expecting to be free to spend their leisure time as they pleased. Some might assist with a few simple household tasks, and help with the digging and planting of the garden, but most expected to get away from home for some part of the evening. On Saturday afternoons they went to football matches and in the evenings to dances or the cinema. Older daughters of large families living at home spent part if not the whole evening helping in the home. On Saturdays some looked after younger brothers and sisters while their mothers went to Boston to do the shopping. They had fewer chances of spending weekday evenings away from home. The demands on the free time of elder daughters of large families could be very exacting. For most their only escape was by going into domestic service and that rarely gave them more free time.

Female attire was most unsatisfactory for farm work. Some local women came to work in strong boots but others, mostly from towns, wore boots and shoes unsuitable for walking over hard clods of earth and shoes gave them no protection during wet weather. In wet seasons dresses and skirts became wet and dirty. The little money women and girls earned went to pay for food and for clothes suitable for wearing when not at work. They could not afford to buy clothes specially for land work. The disabilities under which they worked became less serious when they started to wear shorter dresses and rubber boots but few wore any form of satisfactory protective clothing. Doctors made frequent references to the serious effect which working out in the fields, in all kinds of weather, had on women not properly clothed for the work. Women made no attempt to fit themselves out with clothing similar to that worn by the Women's Land Army during the war. Perhaps it could not be produced commercially at prices which low paid workers could afford.

On farms where several women were employed the common practice was, and still is, for them to work in gangs under a gangmaster. Members of these gangs consider themselves as regular workers on these farms. For them the risk of unemployment during the winter months was less serious than for other gangs employed on a casual basis. The latter, organised by gangmasters rather than by farmers, on occasions spent several weeks on one particular farm, but more often for only a few days. For them continuity of employment depended on their standing with farmers in the area in which they operated. The contractual relationship between these gangmasters and their teams varied. It was generally assumed that the individual members were employees of the particular gangmaster. In a legal sense, however, the gangmaster was the employer only when he took work on a contract basis, that is by the piece. At other times when paid at time rates gangmasters acted as agents of the farmers employing their gangs. The conditions under which gangs of women worked were regulated by the Agricultural Gangs Act, 1867. Under it a distinction was drawn between gangs employed directly by the occupiers of the land on which they worked and those employed directly by a gangmaster. Under the Act an agricultural gang meant a 'body of children, young persons or women, or any of them under the control of a gangmaster'. It laid down that gangs employed by and under the control of gangmasters to work on land of which they were not the occupiers could not be composed of both adult male and female workers. Further, a gang of females could not be employed under a male gangmaster unless a licensed female gangmaster was also present. The licensing of gang masters continued in Lincolnshire until the mid-1960s but for many years little attention had been paid to the restrictions relating to mixed gangs or to the need to have a female gangmaster.

In the 1920s farmers provided transport for town gangs. All kinds of vehicles were used depending partly on the distance between town and farm. Large farmers used their motor lorries, others had light horse drawn vehicles. It was not customary to provide seats, workers sat on the floor of whatever transport was provided. Fortunately speed in those days was comparatively slow and there was little risk of anyone falling off when taking corners or travelling along badly rutted farm roads. Local workers, male and female, provided their own cycles to get to and from work. On occasions they had to travel long distances not just from home to farmstead but also to the place of work which might be a mile or more from the farmstead. At times they might be required to change their place of work two or more times during a day's work. This was done in all sorts of weather, at times over poor farm roads and through muddy fields. No provision was made by farmers for the protection of cycles while their owners were at work. Some regular workers considered employers ought to provide transport from farmstead to the place of work. Women in particular argued that they ought to have the same facilities as those provided for town gangs. Farmers, however, insisted that as local women formed a part of the regular labour on the particular farm they had to be considered differently from casual gangs brought some distance from towns. The normal working day for women was from 8.00 a.m. to 4.00 p.m. with half an hour for lunch. The practice of working a shorter day when at piece work was not favoured by farmers, especially when, as with lifting maincrop potatoes, it was important to complete the work before the bad weather set in.

The same town gang, under a woman gangmaster, was employed on a seasonal basis for several years on our farm. It was my only experience of town gangs. Most of the women, girls and boys came from good lower working class homes. Because of the limited opportunities for employment in Boston most of the gangs from there included young girls and boys who worked on the land while waiting for an opportunity to secure work in a town occupation. At times there was a good deal of bawdy cross chat between members of this gang and local men working with it. I doubt, however, whether the general standard of morality was as bad as one might have suspected from listening to the crudity of their banter with men. Before 1921 I had limited experience of working with women and then only with those who lived in cottages on the farm where I had worked. It was a new and very different experience to be working with more sophisticated, less reserved town girls. They had more opportunities than our village girls for mixing with young men. I suspect they knew that farm lads were afraid of them. Their brazen verbal posturing was, I suspect, part of an act put on for the benefit of finger sucking yokels. They were in a situation in which the restraints of family, friends and neighbours no longer governed their actions. Members of gangs which spent only relatively short periods on any one farm had less reason to be careful not to outrage the susceptibilities of farm communities than others who, each year, worked for long periods on the same farm. Local women, often wives, daughters or sisters of men working on the same farm behaved as one would expect of people living is small rural communities. Conversation between men and women was not always 'chapel talk' and young men and women would be a little freer in their conversation when away from adults.

Before 1921 my brothers and I had no idea how villagers spent their evenings and week-ends. We assumed young people became involved in all sorts of individual and group activities. By moving to within a mile of a village we expected to find a wide choice of leisure activities from which to select according to our individual inclinations. We soon found, however, that young people spent much of their free time wandering aimlessly about the streets. I was surprised at the lack of opportunities for people to involve themselves in socially useful and satisfying leisure pursuits. Perhaps they, like ourselves, had lost any urge they may have had to organise their free time and resources for the advancement of village community life. Various reasons can be given for the poverty of social life in rural areas. The cycle and later motor cycle, the motor car and the introduction of regular daily bus services between villages and towns made it possible for an increasing number of young people to spend their evenings outside their own small communities.

Religious, welfare and sports organisation, instead of bringing people together and having a cohesive influence on village communities had, to a large extent, done the opposite. Church and chapel communities were too small, too wrapped up in their own ethos, to be able or willing to provide the leadership necessary to make villages something more than collections of families. Some would blame class divisions, the absence, within the different status groups, of any desire to become involved in activities outside their own group. It was more obvious among women than male members of villages. Some blame rests with the more influential older members of each of the social groups. Whatever their financial status they could have, had they wished, used their influence within their own groups to introduce social and cultural activities to the whole community. Those with the financial means bore the greater responsibility for the poor provision of leisure activities; they had the means to give a lead in any endeavour to finance desirable improvements. In our village they strongly resisted any kind of development which was suggested. The more articulate members made little effort to bridge the divisions between the social, cultural and religious groups. We were all blameworthy. Had we shown some desire for action leaders would have come forward. If people had wanted tennis clubs, choral societies, libraries and village halls they would have made the effort to get them. Unfortunately most villagers kept themselves to themselves, they were good neighbours only within their own groups. When we moved to Sutterton there was no village hall. Later a small group sought, without success, to persuade the Parish Council to provide one. During my time there the Parish Council did nothing to make the village a pleasanter place to live in, nothing to encourage villagers to take an active interest in improving the social life. Farmer members of the Parish Council had been heard to declare that villages did not require recreation centres, that young people who did a proper day's work had no need of physical exercise in the evenings and at week-ends. In the evenings they needed to rest and home was the best place for that. Later a village hall was built with funds raised by voluntary effort. That would have been all right if the effort had been equal to the need, instead it was a mean structure capable of accommodating only a few of the simplest kinds of games and activities for young and old. It was too small for dances. On the few evenings that I spent there I had to be content with playing dominoes. Outdoor sports in the village were limited to football in the winter and quoits in the summer. These provided activities for the few, for the rest there was nowhere to spend free time outside their homes other than in public houses or walking the streets. Leisure time was unorganised and few people gained positions within the community which enabled them to make useful contributions to village activities.

My brothers and I had previously spent so many years in comparative isolation, and become too self contained to have the urge to be initiators of new ideas about the use of leisure time. We were too afraid to free ourselves from the inhibitions bred out of spending so much time within the family. We had not accustomed ourselves to spending our evenings in public houses and lacked the self confidence to seek, within the village, new ways of occupying our evenings and weekends. My eldest brother was more determined than I was to do something useful. He was interested in machines and spent his evenings, when not working on the farm, helping in a small cycle shop. It was unpaid work but for him the pleasure of doing something, the chance of overhauling a motorcycle, was sufficient reward. I was not interested in machinery and looked forward to Saturday evenings when I could get away to Boston for the evening. Before 1921 I had started spending Saturday evenings in the small town of Donington, about six miles from where we lived. It lacked a purpose built cimema but the Town Hall was hired for a show every Saturday. Each weekly programme included an episode from a serial film. I cannot remember the serial film being shown when I started my weekly visits. I saw all the episodes of *Circus King* in which Eddie Cantor was the principal male actor. We had an occasional breakdown during the show but I don't remember an occasion when a show had to be cancelled. From the start of my weekly visits to Boston my pal and I followed a routine which, so long as the visits lasted, was not varied. We rarely spent 2s. except when we had to buy clothes and other necessities. Our normal Saturday evening expenditure was made up as follows: 1d. for parking our cycles, 4d. for a quarter of a pound of sweets, 7d. for a seat in the cinema, and 6d. for fish and chips, making a total expenditure of 1s. 6d. We had our supper after leaving the cinema, eating it on the street out of newspaper. We complain today about the untidy habits of young people but I am sure there was more rubbish on the streets of Boston on Sunday mornings in my young days than there is today.

Chapter 12.

UNION MEMBERS UNDER PRESSURE

IN A NUMBER OF RESPECTS the conditions for farm workers during the period from 1918 to 1921 were good, certainly much better than in earlier years. The improvement in wages enabled young people, if they wished, to save a little from their weekly wages. Mother had always encouraged us to save. As schoolboys we saved the pennies received from catching rats and moles and for doing other tasks about the farm. With these I bought my first cycle for 12s soon after leaving school. Later when we handled our own wages mother opened banking accounts for us. From then onwards she was constantly urging us to add to the money put by. By 1921 I had saved £90 which seemed a very large sum. It gave me a tremendous sense of independence at a time when I was hopeful of increasing my weekly rate of savings. Mother's brother was held up to us as an example of what could be achieved. As a single man working on farms he had managed to save sufficient to take a County Council smallholding when he married. I don't think my brothers or I had any hope of that kind of achievement but with a good bank balance I hoped to be able to obtain employment outside farming. I did apply, without success, for jobs on the old Great Eastern and Central Railways.

Before we changed farms in 1921 I had a basic wage of 44s and continued to receive that wage under our new employer. I soon discovered that some other lads of my age on that farm had a smaller wage. On occasions during the early summer of 1921 my weekly earnings, including overtime, amounted to 50s or more. This did not satisfy me. I wanted to be free of horsework and to have the opportunity to work with men on piece rates. Although not 18 I was doing the same work as men receiving the adult rate of pay. During the few times when I had a chance to do work by the piece my earnings proved my capacity to achieve the same rate of work output as men. I decided to ask the foreman for the adult rate of pay. At the time of asking we were threshing wheat and the foreman told me I could have 'a man's wage' if I was able to carry eighteen-stone sacks of wheat from the thresher to the granary. I felt bound to accept the challenge. Previously I had carried the odd sack of wheat but never with confidence and doubted whether I could do so for a whole day especially as it meant carrying the sacks up steps into the granary. After doing the work with some difficulty for one day the foreman perhaps realised he had been petty in putting me to the test, at any rate he took me off the work and later agreed to pay me the higher rate. One or two of the men on the farm in receipt of the adult wage of 48s. per week could not do the work so I felt I had justified my claim. By the time I received a man's wage it was in fact what we called an Irishman's rise. The adult wage had been reduced to 42s. My status had gone up — my wage down.

The first half of the 1920s was a difficult period for both farmers and farmworkers. One important result of the improved financial condition of farming prior to 1921 had been a substantial increase in the price of farm land. The high level of prices of farm produce and the guarantees given to farmers by Parliament encouraged them to borrow capital, buy farms, and to invest large sums in land improvement. Under the Corn Production Act, 1917, the control of prices of wheat and oats and of farm wages was due to end in 1922 unless Parliament took action to continue the provisions. This it did in 1920 by passing the Agriculture Act. Unhapplily events obliged the Government, within months of passing the 1920 Act to abolish guaranteed prices and the control of wages.

A serious inflationary situation in this and many other countries forced the Government to take action, which, in conjunction with an alarming increase in imports of cereals and other goods, resulted in a disastrous fall in the general level of prices of farm produce. Wholesale prices in Britain, which in 1920 had risen by 52 points over those of the previous year fell by 110 points in 1921. Prices of farm produce fell in that year from an index of 192 (1911-13 =100) to 119. Continuation of the guaranteed prices for wheat and oats would have involved the Government in paying substantial deficiency payments to farmers. It was decided these would not be justified and the Corn Production (Repeal) Act, 1921, was passed. When the slump came banks and others sought to recover some of the loans to farmers and landowners. Unfortunately many who had borrowed large sums could not satisfy their creditors because of the disastrous fall in the value of land; and many were made bankrupt. There was a general loss of confidence in the future of farming.

Farmers insisted that their industry could not survive under the pressure of cheap imported farm products and demanded protection either in the form of tariffs or the restoration of price guarantees for their cereals. They sought the support of farm workers in their endeavour to persuade the Government to meet their requests. The unions, catering for farm workers, however, were opposed to any form of protection against imports. In their view farmers had forced the wages of farm workers down to an excessively low level in the hope that this would compel the Government to take action. The acreage sown with wheat and oats for the 1921 harvest had been influenced by promises contained in the 1920 Act, this persuaded the Government to include in the Corn Protection (Repeal) Act 1921 provision for the payment of an acreage subsidy of these two crops harvested in 1921. It also recognised that the Agricultural Wages Board and its District Committees had served a useful purpose in bringing farmers and workers together for formal discussions on wages and conditions of employment. It decided that the Corn Production (Repeal) Act 1921 should include the establishment of voluntary Conciliation Committees charged with responsibility for regulating wages on a voluntary basis.

Farm workers in Britain were not unanimous in their condemnation of the abolition of statutory control of wages. The Scottish Farm Servants' Union had never favoured these controls and was pleased to see them disappear. It may be doubted whether this union was voicing the general views of Scottish farm workers. Perhaps it may have been influenced by a fear that under a system of statutory control of wages many farm workers would be more disinclined to join a union. An important factor, however, was its experience of collective bargaining before the passing of the Corn Procuction Act in 1917. In the pre-war years it had succeeded, with the suppport of some farmers' organisations in establishing joint committees in a number of districts. These made recommendations on wages which gained a fair measure of support from farmers and workers in the districts in which they operated. After 1917 some of these committees functioned as statutory District Wages Committees and when statutory regulation was abolished in 1921 some continued operating on a voluntary basis. The Scottish Farm Servants' Union preferred the greater freedom and flexibility of voluntary joint committees, hence their pleasure at seeing the 1917 Act abolished.

The leaders of farm workers in Scotland criticised their colleagues in England and Wales for failing, in the years prior to 1917, to secure, with the co-operation of the National Farmers' Union, the establishment of voluntary joint committees for the determination and policing of farm wages and employment. It was further maintained that after 1917 they had allowed themselves to become too dependent on the Wages Board. In consequence its abolition had left them without any formal machinery other than the Voluntary Conciliation Committees provided under the Corn Production (Repeal) 1921 Act.

In England and Wales most farm workers felt that without a Wages Board it would be impossible, under the serious economic condition of the early 1920s, to hold wages at a reasonable level. It soon became clear to us that we had no effective means of stopping the downward drift in wages, nor of halting an increase in the length of the working week. We lacked an effective organisation to resist the determination of farmers to reduce wages and rid the industry of the Saturday half day. Farmers had only to put a few men on short time and draw the attention of workers to the number of healthy young men walking the roads in search of work. That was enough to discourage men with large families, especially those living in farm cottages, from taking industrial action against any reduction in wages and increases in the lengths of the working week. In a period of less than three years, between 1921 and 1924, wages of adult male workers fell from 48s. to 28s. This reduction of 42 per cent in the weekly wage for a 48 hour week would have been acceptable had the cost of living shown a comparable adjustment. Instead prices of essential household goods fell during this period by only 23 per cent. There was, in our view, no justification for the severe cut in wages since the prices of farm produce fell by only 27 per cent. We knew of the increase in farmers' indebtedness to banks and merchants and felt our wages had been pushed down to below the subsistence level so that more of the farm income could go to pay off farmers' creditors.

The situation was worse than the figures above suggest because of a steady decline in regular employment on farms. Farmers decided in our district that time rates of pay for workers, other than those hired by the year, should be on an hourly basis. This left them free to send workers home at an hour's notice, at any time of the working day. At all times of the year, regular, as well as casual workers, risked being sent home when weather halted work in the fields or when trade held up the marketing of potatoes and vegetables. On our farm few men were offered work under cover on wet days. During winter we could expect to lose some time most weeks. Many who regarded themselves as regular workers on a particular farm had the added risk of being laid off for weeks when work was slack, and with the cut back in work on hedges and ditches winter work was much reduced on most farms. Short period stand-offs occurred when the trade in potatoes and vegetables was slack: longer periods without work resulted from a more general decline in seasonal work. In the two winters of 1922-23 and 1923-24 many married men, females and youths found themselves stood off for varying periods. Farm workers had no unemployment insurance, consequently the loss of earnings during the Christmas season was a particularly serious matter for those whose normal weekly earnings hardly met family needs.

Many wives found themselves unable to feed and clothe their families decently. Some had to seek help from their parish councils. I believe the three years between 1921 and 1924 to have been for farm workers in our part of the country the most difficult in this century. Even those fortunate to have full time employment found it necessary to exercise great care over household expenditure.

We were fortunate. Five members of our family had left school and during the mid 1920s only two, a sister and a brother, experienced long periods without work during winter periods. Other members had short periods of idleness at Christmas time; it was the farmer's Christmas present to us. It was not an easy time for mother, the reduced family income made it more necessary to pay weekly visits to Boston in search of bargains. Getting into debt was considered a dreadful sin. Those fortunate to avoid it often looked down on their less fortunate neighbours. It always seemed odd to me that working class people, with their experience of trying to live within what were for some families inhumanly low incomes, should be so unsympathetic towards others who could not be expected to avoid either getting into debt or suffering illness from under feeding. The self-righteous attitudes of those free of debts was always more in evidence among working class people than among the more fortunate members of village communities. Few working people had accustomed themselves to buying on credit, they did not have monthly accounts which could be settled at irregular intervals. Not all wives of farm workers could go to towns and take advantage of bargain prices in company shops. Those with young children, none old enough to look after the younger ones, with husbands unwilling to stay at home and look after children or to do the shopping, had either to depend on village shops or travelling tradesmen's carts. Families most in need of bargain shopping had to pay the higher price in village shops.

Some smallholders also had a lean time in the 1920s, indeed a few must have had less money to spend on themselves than farm labourers in full time employment, especially those who worked for a great deal of their time at piece work rates. It was common knowledge, in our district, that some small farmers, heavily in debt to large farmers who were also produce merchants, suffered great personal hardship. A common complaint against these farmer-merchants was that they marketed their own produce when prices were good and insisted on moving the produce of farmers in their debt when trade was dull. These debt ridden farmers did not have the freedom to decide when and to whom to sell their produce. It was not surprising that some envied workers who had a fixed working week and a regular, though low, wage.

It seemed to most workers that some of the larger farmers lived extravagantly. We felt it was hypocritical of them to insist that we had to tighten our belts and dress meanly while they, with their wives and children, continued to enjoy the high life of the properous years. I found the contrast between how we and our employers lived intensely irritating. I could not save from my weekly wage as I had done up to 1921. After deducting 7d. per week for health insurance I was left with 27s.5d. during the winter months. In the summer with a longer working week, plus overtime, earnings were a little more but never sufficient to make any addition to the little I had in the bank. I paid mother 15s. a week for board, lodging and laundry, a mean reward for all she had to do. From the balance I tried to put aside 9s-10s. each week but this went on clothes, repairs to cycle and for the purchase and repair of farm tools. Until we moved in 1921 I had not been required to buy my own tools; spades, forks and hoes. But on the farm where we worked after 1921 all the men had their own tools. On occasions when we had no idea which particular tools we would be using we started from home in the morning with hoes, spades and forks tied to the cross bars of our cycles. We could never afford to buy new clothes for work. The packman with his basket of secondhand clothes paid regular visits to farm cottages and we also bought old army uniforms from the Army and Navy Stores in Boston. I hated wearing ex-army clothes; perhaps I was too sensitive. I noticed it was the fashion in the mid 1960s for young men to wear ex-army or ex-R.A.F. overcoats as part of their evening and weekend outfits. It did not please me to do so in the 1920s. It hurt my dignity, my self-esteem, not to be able to visit a tailor and get measured for working clothes as I had done before wages began to fall.

My father joined the National Union of Agricultural and Rural Workers in 1918. When conditions began to worsen in 1922 he decided other members of the family ought to join. I don't think any of us were keen and some must have shown their annoyance at not being consulted before Father paid our entrance fees. At the next branch meeting he told the Secretary that he did not think he would be paying our monthly contributions. The Secretary's reaction was to suggest that involvement with the work of the Union would change our minds. He made it his business to see that we did get involved for he persuaded members of his branch to put me on the branch committee. This was not difficult. I was not at the meeting and few of those who did attend had any wish to be elected to positions which might draw public attention to their membership of the Union. Most considered it advisable to be discrete lest it should encourage their employers to oblige them to leave the Union. Having been appointed a member of the branch committee I soon found myself elected as a delegate to Boston Trades Council.

During the war and immediate post-war years, when labour on farms was scarce, there had been little outright opposition, by farmers, to farm workers belonging to a trade union. But under the changed conditions of rapidly falling prices of farm produce and the growing industrial unemployment farmers could, and some did, use their power to oblige their men to leave the Union. Over the five years from 1919 to 1924 membership of our Union fell from just under 127,000 to just under 29,000. Two-thirds of this loss occurred in 1922. It was due in part to a general feeling, by workers, that membership of the Union put their jobs at risk. The reason most commonly given for discounting membership was that the Union could do little, after the loss of the Wages Board, to protect wage and working conditions. In these circumstances it was not difficult for farmers to persuade their men that Union dues went to pay agitators who spent their time undermining the good relationships which, they asserted, existed between themselves and their men. Some farmers seeking men inserted in press advertisements the words 'no trade unionist need apply', whilst others, who felt thay could be more particular in other matters inserted in their advertisements the further instruction that 'no man over 45 need apply'. In these circumstances I was nervous of becoming too involved in the activities of the Union fearing that with five members of the family in it there would be difficulties for Father.

It was a waste of time for workers to try and justify to farmers their right to belong to a trade union. Indeed few farmers would enter into a discussion on the subject since they had no intention of recognising workers' trade unions and would not knowingly employ any man who was a member. When we had a chance to speak to farmers we admitted that trade unionists and their officials did cause trouble between farmers and workers, but that did not mean trade unions ought to be abolished. Nothing was achieved by reminding them that as members of their trade union they used their collective power for economic and political purposes. When we pointed out to them that farm workers also had group interest which could be most effectively safeguarded by the activities of the establishment of a strong trade union we were told that the interest of farmers and farm workers were inseparable and adequately safeguarded by the activities of the National Farmers' Union. We knew that farmers looked on labour as a factor of production, as an item of farm costs which had to be kept low if the industry was to remain competitive with overseas producers. We also knew that the prime interest of employers was to maintain a high level of profits which, in their view, could be most easily achieved by reducing wages and increasing weekly hours.

No one had any detailed factual knowledge of the economic situation of agriculture in the early 1920s. Only a very small number of farmers kept accounts of their expenses and returns and very few inside or outside Parliament had any accurate information on the general economic conditions of landowning and farming. The most reliable published information related to farm workers. This showed that their standard of living had been forced down to an unreasonably low level. Trade union officials constantly emphasised at meetings that when farmers claimed that farming did not pay it should not be assumed that landowners or farm workers got their fair share of the net returns from farming. Certainly no intelligent farm worker at that time considered his wage as reasonable or as high as the industry could afford. We knew there was a conflict of interest, that without a strong trade union farmers would cut our wages rather than suffer any reduction in their own high standard of living. Farmers continually insisted that falling prices had forced on them the necessity to pay low wages. In discussions at Union meetings we suggested that defective management, that failure to make adjustments in systems of farming to meet prevailing market conditions, had been an important contributory reason for the unsatisfactory state of the industry. When it was suggested to farmers that they needed to remedy faults in management a common reply was to the effect that they did not pay us to advise them on how to run their farms. Too many farmers, in our view, considered that prices and costs ought to yield a reasonable profit when practising their preferred, and often outdated, system of farming and methods of production.

Criticism of trade unions and of their paid officials was not confined to farmers. I suppose every branch of our Union had at one time or another the odd member who was frequently in trouble, who rarely secured employment with the better type of farmer, and who rarely left the Union. Unsatisfactory farmers and workers have, invariably, to suffer each other. We had one member in the 1920s who rarely managed a full year without one or more complaints against an employer. Most of these were trivial, the odd one related to what he considered to be wrongful dismissal and threatened eviction from his cottage. There was little the Union could do for this class of worker. Farmers had little difficulty in getting the required court order for possession of a house, at best the Union might get the date of an eviction delayed and thereby give the man a few more weeks in which to find alternative accommodation.

We could expect a member who was frequently in trouble to be critical of the Union. No official from branch secretaries and local organisers to those at central office ever gave this kind of disgruntled member satisfaction. He was troublesome in a more serious way in that by his vociferous criticism inside and outside branch meetings he undermined the support given to the Union by timid members.

Low wages and an increase in short term unemployment greatly dampened any wish men living in tied cottages might have to continue membership of the Union. Under these pressures an increasing number became more conscious of the fact that they were not free men. The risk of losing employment and home imposed a restraint on the activities of men aware of the need to make a stand against conditions being imposed on them. Understandably they wished to be rid of unwarranted restrictions on their economic and political freedom, on their desire to feel free to play an active part in the labour and trade union movement. Farmers still deny that occupants of farm cottages feel themselves to have less personal freedom than workers living in free houses. There are others who have two standards of rights and justice, and have difficulty in seeing that the rights which they demand for themselves are, in essence, no different from those which their workers considered they should have.

In the National Farmers' Union Journal, towards the end of 1971[*] a letter writer wondered why it should be assumed that the tied cottage had an unwarrantable influence on the freedom and social status of the occupants. He saw the position of workers living in farm cottages similar to that of the Prime Minister and the Chancellor of the Exchequer who occupy official residences for the period of their appointments. It seems incomprehensible that anyone should fail to appreciate that persons of wealth, with private as well as official residences, have not the same sense of

---
*Quoted in the British Farmer and Stockbreeder 13/11/71 p.17

uncertainty and insecurity as that experienced by farm workers living in tied cottages. There is uncertainty about the Prime Minister's tenure of official residences but he has no uncertainty about finding another house after vacating that provided by the State. Occupation of 10 Downing Street does not limit a Prime Minister from speaking his or her mind. The circumstances are vastly different from those of men entirely dependent on their weekly wages and on retaining occupation of their houses. Men afraid to do anything which might displease their employers for fear of being without work and homes, who have little or no hope of finding alternative accommodation, have good reason for feeling that they are not free men. For them eviction may, in the short run, mean going into sub-standard houses or flats. It might mean families having to split up and seek temporary shelter with various relatives, possibly husbands being forced to live several miles from wives and families. It often means having to store household belongings in unsatisfactory buildings. The case for and against tied cottages in agriculture has been a subject of heated argument for generations. Farmers have consistently argued that it is necessary for some workers to live near their place of work. This is less important today than when men had to walk to and from work. Now that they have motor cycles and cars, the distance between home and place of work is not so important as formerly. Indeed it could be argued that distance between home, shops, and schools is much less important than is commonly recognised by farmers and some farmworkers. Wives and children get more consideration than was the case when I was a school lad. More workers prefer to live in or near villages and towns. Few are now willing to live in farm cottages, especially in those attached to isolated farms. It is an advantage to have men, who are in charge of livestock, living reasonably near the farms on which they work. Farmers have found, however, that many of these men will no longer live in farm cottages that are not reasonably near rural communities, and that are not situated near a good metalled road. When I was a child my parents could never be very choosey about the location and condition of the houses they occupied. It is different today. Farmers must now provide cottages comparable to those provided by local authorities if they wish to get good stockmen and foremen. There was, in the 1920s and still is today, a good deal of criticism of the low standards of housing provided on some farms. When I worked on farms many labourers complained of serious defects in their houses. Few, however, showed any keenness to apply for higher rented free houses. Some, with large families, could not afford the high rents of council houses. Others, who could, preferred to sacrifice the convenience and comfort of these houses in order to have the extra cash which low rented farm cottages gave, to spend on other things. I doubt if, in the 1920s, many farm workers favoured a change which would have turned farm cottages into free houses but at higher rents. In fairness it must be said that at the then level of farm wages few could afford to add to their household expenditure by taking a free house. In some instance the unsatisfactory condition of farm cottages enabled their occupants to secure a reduction in the general rents fixed by the Wages Board. Local authorities could require owners of farm cottages to put them in a proper state of repair. Perhaps we had too many landowners and farmers as our representatives on County and Rural District Councils.

Chapter 13

WORKERS LOCKED OUT

IN 1921 THE GOVERNMENT no doubt considered that it had dealt fairly with farm workers by providing Voluntary Conciliation Committees to take the place of the Wages Board and District Wages Committees. It had been agreed that if any of the Conciliation Committees unanimously desired that its decisions should be submitted to the Minister of Agriculture, these would, if confirmed by the minister, become the minimum conditions of any contract of employment in the area to which they related. This had little support from farmers. By March 1923, only sixteen, and by the end of that year only four out of a total of sixty-three Conciliation Committees in England and Wales had asked the Minister of Agriculture to make their agreements minimum conditions of any contract of service.

Farm workers quickly discovered that these Committees gave little protection. The Minister of Agriculture lacked the means for effectively policing agreements In the absence of a large army of inspectors, able to pay visits to farms and inspect wages books, it was left to farm workers to make complaints of underpayment. In the circumstances in which they found themselves at that time few dared to make complaints direct to the Minister and not many asked their union to take action. At the very time when farm workers needed these statutory controls they had been given useless voluntary machinery. We knew that the large number of unemployed urban workers made it impossible to oblige farmers to take any notice of voluntary agreements. In our area there appeared to be a concerted effort by farmers to rid the industry of workers' trade unions. In the absence of a strong effective trade union we felt compelled to take whatever conditions farmers cared to offer.

By the end of 1922 the average weekly wage of ordinary farm labourers in England and Wales had fallen to just over 31 s.

(£1.11s.0d.). It was about a third less than that paid in August 1921. This, however, did not stop farmers from insisting on further reductions. Threats of withdrawal of labour had little influence. In the autumn of 1922 farmers in Norfolk offered their men a wage of 25s. per week and by the beginning of 1923 this was the most common rate paid in that county. Despite this low wage of 6d. per hour farmers decided in February 1923 to reduce it to 5d. and to increase the weekly hours from fifty to fifty-four. Men were asked to work another four hours each week and take a cut of 2s. 6d. per week in their wage. This was rejected by workers' representatives and the offer was increased to $5^1/_2$d. per hour, farmers insisting that this would be imposed. Between 400 and 500 men refused the terms and found themselves locked out.

In our district farmers sought to increase the length of the working week from fifty to fifty-four hours. At first they suggested abolishing the Saturday half day, when this was resisted they increased the length of the normal working day. Men employed by one farming company refused to work the longer day and went on strike. The dispute was settled fairly quickly, the company agreeing to make no change. This incited some workers in our area to propose that a strike be called in support of the men locked out in Norfolk. Union headquarters took immediate action to discourage any extension of the Norfolk dispute. Our leaders recognised such action could have little influence on farmers who knew we lacked the organisation and financial resources to resist their demands. Some of the more militant members of the Union demanded that a determined stand had to be made against the erosion of their standard of living. In their view the Union had to take some positive action otherwise it would suffer further losses in its already small membership. If that happened then all hope of obtaining decent working conditions for land workers would be lost. At branch meetings our officials had the greatest difficulty in restraining some members from proposing measures which could not be sustained by the financial resources of the Union. No one dared to make public the financial weakness of our organisation. Organisers at meetings with members demanding action to improve their wages could do no more than encourage everyone to do their utmost, short of coming out on strike, to resist pressures from farmers to lower wages and increase the length of the working week. It was not surprising that an increasing number saw little purpose in making monthly contributions from their small wage to a union which did so little for them. Few saw themselves as being responsible, through their neglect to encourage more workers to join the Union. Our severest critics were those who had never been members. We could not make them understand that their refusal to join the Union had made certain our inability to secure a reasonable wage, better working conditions and better farm cottages.

The dispute in Norfolk lasted for about four weeks, it ended when the men accepted 6d. per hour and a 54 hour week of which only 50 hours formed the guaranteed week. The rate was 1d. per hour below our wage but we did not have a guaranteed week. Our employer made frequent threats during 1923 to abolish the Saturday half day and in March, 1924 gave instructions that we had to work a full six day week. Although some of the larger farmers in the area had failed to get rid of the half day there had been fairly widespread acceptance of the six day week.

Some employers argued that their action had been prompted by a desire to help families unable to manage on a weekly wage of 28s for a 48 hour week in winter and of 30s.4d. for 52 hours in summer. Working on Saturday afternoon added another 1s.9d. to the weekly wage, a useful addition for those with many small children. This kind consideration, by farmers, ignored two points of view held by workers. First they saw nothing in the pattern of living of their employers to persuade them that the industry could not afford to pay a decent wage for a five and a half day week. Secondly, no one expected the change would result in an increase in the yearly total hours of manual labour expended on farms. Working on a Saturday afternoons would in our view mean either fewer workers being employed or an increase in casualisation and short time working. Many feared that they could just exist on 28s. per week. We feared we would, at the earliest opportunity, be forced to work the longer week for that wage.

In our particular case we felt the instructions to work a six day week had been prompted in part by political animosity. Our constituency had, since 1918, been represented in Parliament by Mr. W. S. Royce, a member of the Labour Party. Previously he had unsuccessfully fought the constituency as a Conservative. Consequently his presence in Parliament as a Labour M.P. was a constant irritant to some farmers. Fortunately most were content to bide their time. They believed the support given to Mr. Royce was personal rather than political, and that the constituency would be won by a Conservative at some future election. Our employer was less rational in his behaviour. At the general elections of 1922 and 1923 a number of farmers had sought to influence the voting of their workers by suggesting that if Labour became the ruling party there would be a general loss of economic confidence by industrialists which would have serious consequences for farmers and farm workers. This failed to swing the majority of voters from supporting Mr. Royce. In the country as a whole, the Conservatives gained, in 1922, the major support but not a majority over all other parties. It was invited to form a Government and continued in office after the 1923 election, when Labour gained the largest number of M.P.s but not a majority over those supporting other parties. The policies pursued by the Government, aimed at stabilising prices, caused considerable financial embarrassment to industry and added to an already serious problem of unemployment. At the beginning of 1924 it was defeated on an important matter of policy and resigned. Ramsay MacDonald was invited to form a Government: for the first time Labour was in office. This was more than our employer could take; for us the day of reckoning was nearer.

Two months passed before he decided to take firm action. In the middle of March the foreman was instructed to tell all workers to work a full day on Saturdays. We understood that reasonable requests for time off would be allowed, a concession intended mainly for two workers who played in the local football team. I did not play football, nor was I an ardent supporter, but I preferred watching the game to working on a Saturday afternoon. Being compelled to work a six day week seemed unreasonable when workers in almost every other industry and occupation had the weekly half day as well as a much shorter working week than our own. The argument that the weekly half day was necessary for workers in shops, offices, and factories, that they required the extra time to get into the fresh air, did not impress us. We got plenty of fresh air when at work, often too much. Some suggested that perhaps an arrangement could be made for farm workers and shop assistants to swap jobs from time to time. We knew that few workers cooped up in shops and offices used their free half day to collect a little more fresh air.

From the start of the longer working week a few younger workers asked for time off on Saturdays, giving all sorts of dubious reasons. I am sure the foreman knew our actions represented a determination to avoid, as often as possible, having to work a full Saturday. In granting our requests he no doubt hoped we would, given time, get used to the idea of treating Saturday as an ordinary weekday. Our employer was less rational, he would have none of the foreman's tactics. He gave instructions that we either worked a six day week or 'collected our cards'. This instruction was given in the middle of a week. On the following Friday when we went to the office for our pay each was asked if he or she intended to work a full day on Saturday. Those who said 'No' received their insurance cards and knew their employment had been terminated. We had, so far as I know, no legal remedy against being sacked at such short notice. Having no written agreement we assumed our employer was free to send us home at any time of the day. I don't remember anyone challenging the farmer's right to terminate our employment temporarily or permanently on such a short notice.

Of the 'twenty odd' regular and casual workers on the farm at the time only six refused to comply with the instruction. On the following Saturday afternoon we had a Union meeting at which the local organiser, instructed by Central Office, declared the dispute a lock-out and offered financial support to both members and non-members of the Union if they withdrew their labour. Following this meeting about a dozen workers became involved. The Union rate of benefit for strikes and lock-outs was 12s per week. Married men, especially those with a number of small children, could not face many weeks without work. Few had any savings. In fact a number of wives were in debt to local tradesmen. At the time of our dismissal we had not expected the Union to become involved. Since it had we hoped that other branches of our Union would be able to give us financial support. Most of us knew that our own branch had no funds of its own and I don't know why we should have thought other branches in the district would have any to pass to us except by way of collections from their members, few workers had any spare cash. Local branches of other trade unions had their financial problems with many of their members unemployed. Most people, believing we had little chance of success, saw little point in giving financial support to us when many more of their own members needed help. We received no support, financial or moral, from local tradesmen and would have been surprised if we had since an increasing number of farm workers and their wives had taken their trade from village shops to company shops in Boston. Our M.P., Mr W. S. Royce, sent two cheques totalling £20.00. The only other help came from a lady in Watford who, after reading of the dispute in the Daily Herald, sent, on two separate occasions, half a crown. She had a good knowledge of the Bible, or at least of those parts which suited her purpose. In her two letters she quoted from it against the capitalist system and Tory politicians. We used the money received to supplement the lock-out pay of married men.

I was appointed treasurer of the lock-out fund. With so little money to hand it was not a very responsible duty. Nevertheless it marked me out as one of a small number of local people considered responsible for the general agitation in favour of better wages and a shorter working week. The dispute occurred during the busiest part of the spring season; this, we hoped, would oblige our employer to take us back on our terms. Our action, however, caused him little inconvenience. He had no difficulty in getting all the labour needed for the most urgent tasks. Young unemployed workers who had exhausted their unemployment benefit roamed the country accepting any work they could obtain. There was no way in which we could stop them from taking work on our farm. In response to a request made to Boston Labour Exchange some unemployed men received green cards directing them to our farm. The issue of these cards was discontinued when the Union notified the manager of the Exchange that an industrial dispute existed on the farm. Boston branch of the Unemployed Workers' Committee Movement did give us assistance by discouraging unemployed workers in Boston from seeking work with our employer. They also demonstrated their support by marching the six miles from Boston and burying a collection of green cards in front of the farmer's house.

In the particular circumstances of the dispute and lack of funds the Union had little choice but to advise us to find other employment on farms where we would not be required to work on Saturday afternoons, except by way of necessary overtime. We knew that, sooner or later, we would have to return to work on the farmer's terms or find work elsewhere. One of my brothers and myself learned that we had no choice but to seek work on other farms. Fortunately no attempt was made to involve workers on other farms in the dispute. By the end of the second week some of the men had returned to their jobs and by the end of the fourth week all except my brother and myself had either gone back to work or found employment on other farms. Father had returned to work having no alternative if he wished to remain in the farm cottage. On returning to work he was told by the farmer to tell 'those two boys of yours to put their feet under someone else's table'. It was a clear instruction to Father to see that we left the district.

Both sides to the dispute behaved rather foolishly since it was reasonably certain the Labour Government would succeed in its intention to re-introduce statutory regulation of farm wages and hours. Our employer must have known that his insistence on a six day week would be short lived, it was a victory not worth the upset. On our side we knew we had no hope of resisting the temporary loss of the half day. It was a pity, therefore, that we did not stifle our irritation, accept the temporary loss of the half day, and await the passing of legislation to control wages and hours.

At first my brother and I had no intention of leaving home but as the days passed and no farmer within reasonable distance from home seemed in need of workers we decided we had no choice but to move out of the area. I suspect that some of the farmers whom we approached declined to offer us work because of the rumours that we were trouble makers. Bringing the Union into the dispute only made local farmers less willing to offer us work. After six weeks without work we attended the hiring fairs held in May intending to take work as horsemen. Boston May hiring was on the Wednesday following May 14th. When we arrived there that morning we found an exceptionally large number of men already in Pump Square. where men seeking yearly hirings foregathered. The decline in prices of farm produce and their effect on the profitableness of farming had persuaded many farmers to make changes in their system of farming. This had reduced the number of workers in full-time employment and persuaded a larger than usual number of horsemen not to change their employer. But an increasing number of other single men, who in more favourable circumstances would have sought work as day labourers, decided to seek employment as horsemen. By taking yearly engagements they hoped to avoid the risk of irregular employment, or perhaps of long periods of unemployment. These and other problems resulted in an unusually large number of men competing for the smaller number of vacancies.

Seeing this large number of men in the Square that morning my brother and I feared we had little chance of being hired. We knew farmers would ask for information on our practical experiences and also where we had previously worked. This information we felt would discourage any from making an offer. I am fairly certain that had it not been for the intervention of our maternal grandmother neither my brother not I would have been hired that day. She was a very determined lady and a foreman of her acquaintance was needing a hired man for one of the farms under his management. Grandmother introduced both of us to him. I suspect he knew all about our previous activities. Perhaps grandmother had persuaded him to overlook what she regarded as our youthful indiscretions. At any rate he did not ask any of the awkward questions. My brother, not quite a year younger than myself, was hired for £50 per year plus board and lodgings.

For reasons which I will mention later, my brother and I had agreed to stand together in the Square and if it could be arranged he would have the first chance of any offer that might be made. After he had been hired he had no wish to stay longer in the Square and I could not face being there on my own. We spent the rest of the day in the fun fair but having little money had to be satisfied with watching rather than participating in the fun. I got no pleasure from the hiring fair that year, the last I attended. I returned home with mixed feelings, my future seemed so uncertain. Had an offer been made I would have taken it even though I hated the prospect of being a horseman. A week after the fair I managed to find work on a farm, not too far from home, where the men still worked a five-and-a-half-day week. I had been told a man was leaving this farm and made it my business to be there asking for work after he had handed in his notice. I said I was prepared to work with horses! This may have influenced the foreman. After all the upset my brother had, by his yearly engagement, put his feet under someone else's table; I continued to live at home.

Before the end of 1924 the Labour Government had re-established the Wages Board and District Wages Committees. The first Order for our area fixed the minimum rate of wages of ordinary labourers over 21 at £1.16s for a 54 hour week. This was 5s above the common rate paid before the Order became effective. We had been told on numerous occasions that at the prevailing prices of farm produce farmers could not afford to pay more than £1.11s, that any increase on that figure would result in farmers changing to less intensive systems of farming. We did not believe that those on the rich silt lands would, as they had said, turn to stick and dog farming (grassland sheep farming). Nor did we think it would result in an alarming increase in casualisation of farm labour. We knew farmers had reduced their normal labour requirements to the minimum. In our view the obligation imposed on farmers by District Wages Committees would benefit the industry. It would require farmers to pay closer attention to management. We hoped it would force them to seek ways of reducing the number of middlemen who, we considered, took too great a proportion of the price paid by consumers for their food. But since many of our large farmers were also produce merchants there seemed little hope that they would tackle the problem of middlemens' margins. Some of us hoped higher wages would oblige farmers to cut down on their own and their family expenditures.

During the dispute in April I had been introduced to a lad who after working for a few years on farms in another part of the county, gained one of the junior scholarships offered by the Ministry of Agriculture to sons and daughters of farm workers and of other parents of comparable economic status who worked in agriculture and related occupations. The Ministry offered, each year, 150 of these scholarships which enabled successful applicants to spend a year at a farm institute. In addition, ten senior scholarships enabled winners to take degree or diploma courses at agricultural colleges and universities. When I met this lad he was a student at Moulton Farm Institute, Northamptonshire. From him I learned that the scholarships were granted on the results of an interview. If applicants had been required to sit a competitive examination I doubt whether I would have applied for one. Having been eight years away from school I knew my chances of winning one by a formal educational test would be small. Applications had to be in by the end of April which left me with little time to ponder on my chances of success. This was fortunate, having decided my application was away in the post before I had time to change my mind. This decision had been the main reason why I had agreed that my brother should have the first offer of an engagement at the hiring fairs.

Chapter 14.

The First Escape

ALTHOUGH ANXIOUS TO GET AWAY from agriculture I had made no sustained effort to improve my chances of obtaining other employment by attending evening classes. For five years after leaving school we had lived too far from any town to be able to attend such classes had I known of their provision. After moving to Sutterton and becoming involved with the work of our Union I learned of evening classes provided in Boston by the Local Education Committee and by the W.E.A.

After becoming a delegate to Boston Trades Council I realised the need to improve my education. Other delegates, especially those from the craft unions, seemed to have a wider knowledge of economics, politics, and social history. I wished I had their capacity for self expression, their confidence that they knew the facts and the conclusions to be drawn from them. I was not competent to make rapid analysis of verbal statements, to decide, at the instant, whether on the basis of logic or history the statements were true or false. Because of my inadequate knowledge on economic and social problems I was afraid to express my opinions after hearing other timid delegates, who did join in the discussion, being savaged by more assertive members. I decided I had better attend evening classes but found that only English, commercial, and technical subjects were provided by the Local Education Authority. These might help me to get work more to my liking but did not satisfy my wish to learn more about the history of working class movements. Later a group of the younger members of our branch of the Union persuaded a headmaster of Kirton elementary school to give us some instruction in English. I found the instruction interesting but not what I wanted most of all. When I learned that the W.E.A. organised classes in Boston I made enquiries and found the subject at that time was English

Literature. By then I had been granted a Ministry of Agriculture scholarship.

In my application for a scholarship I had expressed a wish, if successful, to spend a year at Kirton Agricultural Institute which was about three miles from my home. In consequence I was interviewed by a local committee on which our Union was represented. I cannot remember much about the interview except that I was asked for details of my formal education and experience as a farm worker. A few weeks after the interview I was informed that I had been granted a scholarship. At the time it seemed incredible that I should receive more to go to 'school' than the weekly wage paid to a skilled farmworker. I tried to convince myself that it had been gained on merit, but suspected the real reason was a lack of applicants wishing to take courses of instruction at the newly established Kirton Institute. Previously it had been a grammar school. As a Farm Institute it accepted its first thirteen students in October 1924. I imagine the number was much smaller than had been anticipated. It was certainly smaller than was desirable. It served a comparatively rich agricultural area with many large farmers whom, one might have expected, would send their sons to the Institute for training. Perhaps they preferred to wait until it had demonstrated its value as a training centre. Unlike other farm institutes the instruction given at Kirton excluded practical work on the Institute farm. I did not know this when making my application but was pleased. I had, I thought, done more than enough practical work on farms. I don't know how much farm work other stundents had done. Five, like myself, had gained Ministry of Agriculture scholarships and I assume they had some practical experience of work on farms or in gardens. The other seven, sons of farmers, nurserymen and corn merchants, came to the Institute from schools. I doubt if any of them had spent any length of time working alongside farm workers.

The opening ceremony was performed by Sir Daniel Hall, Scientific Adviser to the Minister of Agriculture. Before that he had been first, Principal of the South-Eastern Agricultural College

(Wye College) and then Director of Rothemstead Experimental Station. As one would expect Sir Daniel was a strong advocate of the importance of a good agricultural education for farmers and their workers. Without this farmers could not, in his view, take full advantage of the training provided at universities and agricultural colleges. This, he believed, had been an important cause of the failure of many landowners to provide leadership in improved methods of agricultural production and marketing. He also emphasised the need to encourage young farm workers to take the training provided at farm institutes. In his speech at the opening ceremony he dealt in some detail with the importance of agricultural experiments and demonstrations to farmers in South Lincolnshire. At the end of his address one of the largest farmers in the county, Ald. H. P. Carter, in seconding the vote of thanks, said (quoting from the local press), it would 'please him very much, as an old member of the County Council, if they could realise that this Institute could be run on practical lines and show a profit at the end of the year'. It was clear that Ald. Carter favoured using the Institute farm as a commercial undertaking rather than for experiments. Sir Daniel, in a long reply to the vote of thanks, told his audience that the Institute farm could not be an experimental farm if it was required to produce a profit. He explained that: 'the farm attached to a college was often wanted first to show students a piece of good sound farming, but then they were just teaching the routine of a particular piece of farming. In such cases the farm "ought to be made to pay, and ought to aim at paying" because the intention was to demonstrate good practical management. But the purpose of the farm attached to Kirton Agricultural Institute was for experiments, "to find out knowledge".' 'If' he said, 'Ald. Carter tried to run plots of his farm experimentally he would find out how much they cost and how much they upset the routine of farming . . . they could have one or other, but if they did experiments they must make up their minds it would cost money. They could not have them (experiments) and at the same time have the ordinary farmer's balance sheet'. In effect he told them the farm could be made to yield valuable scientific and technical information or show farmers good profitable farming practices on that type of land. He suggested they would find the information gained from experiments more valuable than any cash profit which might be had from running the farm as an ordinary commercial business, that any financial losses which might result from experimental and demonstrational work could be more than offset by the higher profits which individual farmers obtained by using the knowledge gained from work done on the Institute farm. And if they required the Principal to run the farm as a commercial business they would be setting him an easier task than that involved in planning and managing experiments.

Eight of the students, including myself, lived at home, the other five in lodgings in Kirton. Consequently there was no student community life outside the classroom. In any case the number was much too small to justify any serious attempt being made to encourage student activities during evenings and other free periods. Living at home was, in my case, a great disadvantage. With our large family and small house it was impossible to have the facilities necessary for quiet study in the evenings. I had either to study in the living room which, during the winter evenings was always crowded and smoky or return to the Institute and work in the library. A hot smoky room dulled the senses and trying to read in a room with only one paraffin lamp soon became wearisome. It was never suggested that I should use our front room: my parents would not think it right to make special provision for one member of the family. We had no favourites in the family. The weekly payment made to mother by the authorities for my board and lodgings was higher than what I had previously paid and more than those made by other members of the family. This, however, did not seem to my parents a good reason for making any distinction in their treatment.

It was not appreciated that, as a student, I required facilities for studying at home. In later years when I came home on vacation from college I was unable to do serious study at home. I suppose my parents thought students, like farm workers, did all their work during a normal working day. Home was not a place where anyone other than mother worked. As school children we had never had homework and as I have mentioned earlier private reading was never actively encouraged. Most rural working class families at that time considered education to be the responsibility of schools and other educational institutions, the pursuit of formal education began and ended inside school or college. The other five students with scholarships were more fortunate. The authorities, in selecting lodgings for them, made sure that adequate provision was available for private study in the evenings.

Looking back I am sure I would have gained more from spending a year at a farm institute with hostel accommodation. Apart from the better facilities for private study there was the enormously important value of living with a number of students having widely varying experiences and backgrounds. It was what I needed to give me greater confidence and to improve my speech. I doubt if any of the other thirteen students that year lacked to quite the same extent as myself the ability to mix with people outside their own narrow social class. During my first weeks at the Institute I found myself looking on some of the other students as people from whom I took orders. It was not easy to adjust to a situation in which young men from different socio-economic groups spent their days working together as equals. The fault was mine, my inferiority compelled me to stand off and miss much of the little of what that year had to give. My fellow students were very friendly but it was difficult for them as for myself to find common ground for social intercourse. Difference between our social backgrounds, home conditions and spending habits were as obvious to them as to myself. My past experiences, education and social life limited any contributions which I could make to discussion inside or outside the classrooms. The five scholarship students whose home conditions may not have been greatly different from my own went to their lodgings for lunch while I ate my sandwiches with the other seven students and discussions were mostly on topics outside my experiences. In so many ways I saw myself as the odd one. Had there been more students my different training, social habits and experiences might not have been so obvious. Had students been drawn in larger numbers from a wider range of farmers and workers it would have been easier for the shy individual, too concious of his own limitations, to hide himself in the crowd or better still to find his own niche in a more diversified community.

When the weather allowed I spent my evenings in the Institute library. Conditions for working were much better there than at home, but I did not find it easy to study. Having previously done so little reading, especially on serious matters concerning the world around me, I had no idea how to begin the task. It took me some time to realise that one did not read textbooks as one did light novels. Because of my lack of any previous training in science I was often at a loss to understand the explanations and arguments of scientists and had the greatest difficulty in making myself stick to the task. In the library I felt that I ought to be able to concentrate. When studying at home I could blame the hot smoky atmosphere of our kitchen for wanting to sleep. It was not much better in the library, the warm room and strong artificial light soon made me drowsy. Another problem was that of learning the language of scientists. Many scientific words were meaningless to me and definitions did not advance my understanding of them. In the lecture room I had difficulty not only with these scientific words but also in trying to take intelligent notes. I don't think that the lecturers had any experience of the needs of students with only an elementary education. Most students at farm institutes required hand-outs giving the main points of each lecture and advice on supplementary reading. I gained the impression that perhaps one of the thirteen students in my year at Kirton fully understood the lectures given by the agricultural chemist. Lecturers were placed in a difficult situation, had they sought to satisfy the needs of students lacking a grammar school education the time of other students would have been wasted. They did their best in the circumstances but it meant that I did not get the help required in the early weeks of the courses.

Trying to take full notes of lectures was not a task beyond the capacity of the best class of student. For me it was impossible to take any which could be used later. In the past I had done so little writing consequently my speed was slow. I knew nothing about the art of note taking. I could not spell many of the words of ordinary speech and in the lectures was soon lost in a jungle of scientific jargon and Latin words. Before long I gave up trying to take notes, I decided for good or ill to rely on textbooks and laboratory work and on asking questions. The latter I did outside the classroom, not wishing to exhibit my ignorance to other students. I enjoyed the laboratory work and learned more of the subject matter of courses from this than from lectures. However, it was not easy to find the right scientific words to describe what I had learned. I was also slow in performing the work, clumsy when cutting sections from plants for microscopic examination. I could handle farm tools better than laboratory instruments, a cabbage knife better than a scalpel.

At the end of the courses we sat examinations, something I had not done for nine years and then nothing comparable with what was now required. Because of my inadequate knowledge of the sciences I dreaded having to compete with lads from grammar schools. They had, in addition to their training, experience of taking examinations; they knew how to organise the time allowed for each examination. I feared I would not have at hand the right words to state what I knew, or thought I knew, on the different subjects. It was, therefore, a pleasant surprise to learn that I had gained a second class certificate. In my case the grade of certificate was important as a passport to a position in agriculture better than that of a farm labourer. It was not so important to students who, after completing the training, would be going back to positions of responsibility on their fathers' farms. Some of these students neglected to take the examinations seriously and this was, I believed, the reason why they gained lower grade certificates than I did.

I had no idea what kind of employment one could expect to get at the end or a year's training at Kirton. Mine had been mainly concerned with arable crops and was more suited to the needs of sons of local farmers than to students requiring technical training in all the main branches of crop and livestock husbandry. This wider training would have increase my chances or securing a position of some responsibility in any one of a number of different farming type areas. The lack of any detailed training in milk, beef, and pig production meant I had no hope of getting employment on dairy or other livestock farms other than that of a general labourer. The training at Kirton Agricultural Institute was, I soon discovered, too parochial. Perhaps those responsible for planning the courses considered they could not deal adequately with each of the main farm enterprises in Britain. It may have been that the County Council considered the Institute's first, perhaps its only responsibility, was to serve the needs of farmers in its own county. But students who came from North Lincolnshire and one who had, before taking the courses at Kirton, worked in a private garden, must have felt that more attention ought to have been given to their particular needs.

If it had been possible for the Institute to attract a much larger number of students from widely differing farming backgrounds my time at the Institute would have been of considerably greater value to me. Exchanges between students with this wider background of farming experience would have given me a greater awareness of the complex nature of British farming and of the problems of production and marketing. With a larger number of students from a wider area I would have gained a better understanding of other aspects of rural, social and cultural problems.

Kirton Agricultural Institute lacked the staff and equipment to satisfy the needs of students seeking training in systems of farming in other areas with differing soil and climatic conditions. The professional staff of five, including the Principal, had, in addition to their teaching duties inside and outside the Institute, responsibility for experimental, demonstational and advisory work in the county. It was impossible for them to give even the minimum time which would have been necessary if the training had included animal husbandry. The small number of students in that first year did not justify the number of staff required to give instruction in every branch of farming. At the time, however, I was not aware that the restricted range of courses might prove a disadvantage when looking for employment at the end of the course. I had been too deeply involved in trying to cope with lectures and laboratory work to have time to consider whether the training would improve my chances of gaining a position of responsibility in agriculture. Indeed when I started the courses I could not claim to know why I had gone to the Institute. I applied for the scholarship because I was out of work.

In the circumstances I considered myself fortunate to be offered a temporary summer job at the Institute after completing my training. I worked on experimental plots and assisted with laboratory work. The experience gained from this work was most valuable; it gave me a chance to learn, in a more satisfactory way than lectures had done, the importance of a proper understanding of the science of crop husbandry. It was a more leisurely way of gaining knowledge and suited my sluggish mind. The work also gave me some training in the skills required by technical assistants, in particular the importance of careful recording of physical data and other observable facts necessary for a proper evaluation of the results of experiments.

This work was temporary; perhaps I might have been offered a permanent post as technical assistant since the Institute had not then obtained all the technical assistants it required, but as far as I knew my work was to finish at the end of the summer. Consequently I was very pleased when Mr. H. W. Miles, entomologist at the Institute, offered me a post as his technical assistant when he left to take up an appointment with an American firm which was introducing the pesticide Cynogas into this country.

During the summer I had done some work for Mr. Miles which must have given satisfaction for he seemed happy to have me as his assistant. One of his particular interests, in his new post, was the destruction of wireworms and my first task was to make a count of the wireworm population in a newly ploughed-up grassfield near Boston. In addition to collecting data for estimating the number of wireworms to the acre I also recorded information of the influence of changes in soil temperature on the movement of this pest. I worked on this project during the autumn of 1925 and then took charge of a series of experiments of ways of attracting wireworms to the upper layers of the soil in sufficient numbers to justify using Cynogas to destroy them. I worked with nurserymen in the home counties, South Mimms, and on the south coast and with strawberry growers in the Isle of Ely. The aim was to obtain information on the best bait and temperature to attract wireworms to within three inches of the surface. These experiments and demonstrations gave me an opportunity to travel outside my own county and to mix with people having widely varying experiences. Staying at hotels favoured by commercial travellers put me in touch with people who exuded confidence in themselves and their products. It was the kind of experience I needed but I cannot say that it did a great deal towards improving my own self confidence. I did not dare to compete with them; they knew too much about everything and as I knew so little I deemed it best to keep quiet.

I worked with Mr. Miles for about seven months until the General Strike in May, 1926, interrupted my work. It was impossible to travel to the various experiments and demonstrations and consequently my work lost any value it might have had and had to be abandoned. The American firm had previously been planning to hand over the main agency for the distribution of their products in this country to a London firm. It also decided that all future experimental and demonstrational work could be left to agricultural experimental centres and agricultural departments of the universities. I was given a fortnight's pay and my appointment terminated. The strike was not the cause of my loss of an interesting job but it persuaded the firm to come to a decision sooner than it might otherwise have done.

Chapter 15.

CHAUFFEUR/GARDENER — BACK TO THE LAND!

THIS LOSS OF WHAT WAS FOR ME interesting work came at a time when industry generally was affected by the General Strike. There seemed little prospect of finding work outside agriculture. Even before the strike some eleven per cent of insured workers, excluding farm workers, were unemployed. The proportion was higher in many parts of the country and a further general increase in long term employment resulted from disruptions caused by the strike. Farmers could get all the labour they required and in the circumstances it was fortunate that the wages and weekly hours of farm workers were controlled by statute. I had no wish to return to the work of an agricultural labourer and had no idea how to set about finding other work for which my training and practical experience might have equipped me. It did not occur to me to seek the advice of the Principal of Kirton Agricultural Institute who, one would have expected, would have been the best person to give advice and assistance.

As I had only paid unemployment insurance for a few months I was able to draw unemployment benefit of 18s. per week for only a very short time. At first this did not worry me unduly because I felt sure the Employment Exchange would soon provide me with information on vacancies that would suit me. I think the officials there knew I would have to go back to farm work. There was little scope in Boston and district for the kind of employment I wished to get. Moreover with so many unemployed persons having better technical qualifications than myself I saw little chance of securing any one of the small number of posts advertised from time to time in the agricultural and technical press.

After exhausting my unemployment benefit I could not wait around hoping work would turn up, I had to get employment, not wishing to exhaust completely my small savings. Many kind, as well as unkind, friends told me that beggars could not be choosers and at last I was forced to agree with them. Although not able to drive a car I applied in June 1926 for a post of chauffeur-gardener. When interviewed I was told I would be taught to drive the car and I accepted the job at a weekly cash wage plus board and lodgings which, in total, was equivalent to the minimum wage of ordinary farm labourers. But with no opportunities for overtime work or piece work I was not so well paid as a farm labourer. It would have been more sensible to have sought farm work but to do so would, I felt, mark me as a failure in the eyes of old workmates. I knew that some of them saw me as someone trying to get above himself. They predicted I would, sooner or later, drift back to the old routine. Some people can accept setbacks with reasonable composure, I could not. Overwhelmed by what a small number of irrational people might think I took the job of chauffeur-gardener, it sounded better than farm labourer.

My employer, a commercial traveller for one of the breweries, used all his slick sales talk when he offered me the post. During my stay there he made no mention of the promise to teach me to drive the car. Since I showed myself able to push it out of the garage for washing perhaps there seemed little point in running the risk of getting it damaged through teaching me to drive. He was a keen fisherman and at the interview mention was made of occasional days out fishing in local drains and rivers. Mention was also made of other opportunities for leisure activities. He had a large garden and a lawn large enough for playing bowls and tennis. Looking at the neglected garden I wondered how I would manage to make and keep it neat if, as had been hinted, some part of my time sould be occupied with fishing and playing bowls. I soon found that the garden and other tasks kept me fully occupied during normal working hours. No one gave me any instructions on what I was required to do in the garden; I saw what was needed and tackled the jobs in my own way.

I soon found that I had little free time from getting up in the morning to going to bed at night. On some occasions my employer, when at home for the evening, had the urge to do a little gardening, and expected that I would be equally keen to work. Living with the family was a disadvantage, it was impossible to get away from work unless I spent my free time away from the house. Although usually free to do as I pleased after tea, especially when my employer was away for the night or when they had visitors, I could never stay around the house and feel that I was a free man, not a lackey.

When they had visitors I still had my meals with them and was treated much the same as at other times. It was the silly little things that irritated me. For example on one occasion some relatives came to stay for a few days and a whole home-cured ham was cooked. When my employer came to carve it, it was clear that he had no idea how professionals tackled the job. He started carving at the fat corner end and seemed uncertain what to do with the first slices because they had little lean meat. He could not offer them to the visitors and was reluctant to put them on the side of the dish. So he passed them to me with the comment 'You like fat, don't you Joe?' It was true that I did not dislike home cured fat ham but that did not seem to me a good reason why I should take what could not be offered to other guests. I was annoyed at being placed in a position where I could not refuse the offer. Temperamentally I was not suited to jobs which tied me so closely to my employer. I could not submit easily to a situation with such imprecise contractual obligations. There was no deliberate attempt by husband or wife to make me feel I was a servant. When they wished me to do something it was always a request. But I could not bring myself to fit into the kind of situation they sought to establish. I had a strong preference for the commonly accepted relationship between masters and servants with a clear undertaking about hours of employment and payment for overtime. It might be considered unreasonable to expect such a formal relationship in the particular circumstances. It was my stubborn resistance to any return to conditions similar to those when, as a schoolboy, I had been at the beck and call of father's employer.

I had to get away. My decision to leave came one evening in

August 1926 when I had arranged to go home and my employer wished to do some work in the garden which required my assistance. When he heard I had other plans he was quite upset. I stayed to help with the work but later gave a week's notice. I was there only seven weeks. Some might think I hardly gave myself or my employer time to get to know each other. At the time I was very unsettled, othewise I might have stayed until I had found other work more to my liking. As it was I put myself out of work, with no unemployment pay and little hope of finding work even though it was harvest time. I searched around without success and feared that if I could not find casual work then there was little hope of securing permanent employment to take me through the winter. I was very depressed and must have appeared so to many of my friends, and in particular to Frank Maidens, a railway linesman who came to my aid. Railway men in many parts of Britain gave considerable help to farm workers seeking to establish a strong trade union and better working conditions. In those days many branches of our Union owed their survival to the support and encouragement given by members of other trade unions. I came in contact with Frank when we moved to Sutterton in 1921. He was an active member of his own branch of the N.U.R. as well as a paid up member of our branch of the Farm Workers Union. He attended our monthly meetings and continued to do so later when he could no longer afford to be a paid member of two trade unions. Frank was an avid reader of pamphlets which supported his Marxist views on socio-economic and political questions. His analysis of working class situations was given in simple language understandable to rural people. Listening to him I was constantly reminded of Owen, the principal character in Robert Tressal's *The Ragged Trousered Philanthropist.* Frank had read the book and obviously appreciated the simple and direct way in which Owen lectured his fellow workers during meal breaks. He had also read many of the books by Jack London and he taught me a great deal about the causes and consequences of our economic and social problems. His understanding of these may have been faulty, that was not so important, what mattered was the encouragement he gave me to find out for myself the things that affected the lives and aspirations of working class families. His explanations of the reasons for the distress and despair of working people avoided the dry logic of professional economists; they were more meaningful to simple people. What he had to say related to the actions of real human beings not the abstract men of text-books on economics.

When the General Strike collapsed in 1926 many workers failed to get their jobs back and others had to be content with working on a part time basis. Railway workers in our village who had to return to their work on a part time basis endeavoured to supplement their earnings by seeking seasonal work on farms. Frank Maidens had been very active during the strike and in consequence had difficulty in getting work on farms in the village. A farmer in a neighbouring village, whose land bordered the length of line on which Frank worked, did offer him work on the days when he was free. I happened to meet Frank a few days after the offer of work had been made and must have looked so depressed that he suggested I should take the harvest work. I knew he could not afford to forgo the chance of extra earnings but he would not accept my refusal to take the work. Being in need of some activity that would take care of my despondency I was not so strong in my refusal as I ought to have been. When I went to see the farmer he seemed, at first, reluctant to give me the work, my impression was that he had not really been in need of extra men but had made the offer to Frank out of kindness. It is also possible that his hesitancy was caused by fear that I was trying to take the job from Frank since he only had my word that Frank had sent me. After a while he told me I could start work the following Monday if it was not raining. He may have had doubts about my ability to do heavy work but I had none and was glad of the opportunity to do something useful. I hoped that when Monday came the weather would allow me to start work before the farmer changed his mind. No matter how dreary the work, the important thing for me was that it put money in my pocket and went some way in helping me to regain my self respect. A few days after starting work I gave practical evidence of my ability to do farm work. One of my first tasks there had been to assist with threshing peas. Later when the nineteen stone sacks of peas had to be loaded on a wagon the horseman happened to have trouble with his back and I was asked to carry the sacks into the wagon. I must admit to being glad when the last one had been dropped in the wagon. Despite the passing of the Corn Sales Act, 1922, which made the hundredweight (cwt., or 112 lb) the standard measure for the sale of grain, farmers continued to sell by the quarter (qtr.). This was originally a volume measure consisting of two sacks each of four bushels. It had, however, become the custom to give each sack a weight measure which varied according to the grain from twelve stones for oats to nineteen stones for peas and beans. Carrying the heavier sacks of grain over rough ground and up granary steps was beyond the capacity of many men and was a cause of injuries suffered by some. I never doubted that the horseman did have a bad back but it was the kind of excuse some might have used in order to test my ability to do the work. It was after this incident that the horseman told me of a conversation which he had with the farmer on the day I was offered work. After making the offer the farmer went to the horseman and said he had given a man a job but did not 'think he was much good for hard work'. He thought I was 'more suited to pushing a pen than a hoe'. This comment was understandable. For some years men from industrial areas had been walking the roads seeking work. After years without work, and with little food, few of these gave the appearance of being able to do heavy manual work. One could easily distinguish between farm workers in search of employment and men who had been unemployed for several months and in some cases for some years. I suppose I looked incapable of standing up to the hard physical work of men during the grain harvest. I was glad to have had the chance to demonstrate that I could do the work. Although my work on this farm was temporary I stayed on for a few weeks after the completion of the grain and potato harvests. When the farmer could offer me no further employment he placed me on another farm which, as an executor, he managed. I stayed there until October, 1927, when I managed to escape for good from being a farm labourer.

I had not given up my trade union and political activities and since the autumn of 1924 had been one of the area secretaries of the Divisional Labour Party. In July, 1926, I was elected to the Executive Committee of the party. It held its meetings on Saturday afternoons. In order to attend these I had to leave work before 1.00 p.m., the normal time for stopping work on Saturdays. No objections were raised to my requests to leave early to attend these meetings and I suffered no reductions in wages for the time lost. The farmer said he had been a Liberal all his life but often added that had he been a 'labouring man' he expected he would have voted Labour. He was, I suspect, a Liberal in the old radical tradition and had he been younger might have voted Labour. But as he often told me he had been a Liberal so long that he could not change. He was the first employer to whom I felt I could talk about political parties and their policies. He was critical of those socialists who were for ever demanding nationalisation of the means of production. But he insisted that workers had a right to be as free as himself to express views on political issues. Whenever we had discussions on political matters he invariably finished up by saying he did not mind what my politics were so long as I did my work. He knew I was not happy working as a farm labourer and encouraged me to apply for posts of higher responsibility in agriculture. But most of those advertised demanded more practical experience of livestock and of technical and managerial skills than I could offer. I could do the technical work connected with experimental plots and field trials, I was also confident about doing the work of junior laboratory assistants. But I was too old for the latter type of post and the wages offered seemed too low for one of my age. There was little chance of getting a post as a field assistant at any of the agricultural colleges, farm institutes or experimental stations. Most vacancies at these centres went to local people. As regards general farming I doubted my ability to take responsibility for planning farm work and controlling men.

I did apply for a post of nursery foreman advertised by Elsoms of Spalding, a local seed merchant. At the interview I was asked about my previous employment and experience. The latter was hardly a recommendation for the post of nursery foreman and this was, no doubt, the reason for my failure to be appointed. But at the time I was sure the troubles of 1924 had caught up with me again. I gave the names of previous employers, This led my interviewer to ask if I had been one of the workers who had refused to work on Saturday afternoons. When I admitted that I had he said he supposed the same thing would happen if I worked for them. I explained there had been no refusal by myself, or any other worker involved in the 1924 dispute, to work necessary overtime on Saturday afternoons or at any other time of the week. I explained that on ordinary farms someone had to attend to livestock at weekends and that I assumed that the foreman of a nursery would be responsible for ensuring that necessary work was attended to on Saturday afternoons and on Sundays. I suggested that there was an important difference between a normal six day week and one of five and a half days. Special terms of engagement for particular classes of workers had to be negotiated by reference to the nature of the work and responsibilities attached to it. All we had sought to do in 1924 had been to retain the five and a half day week. We had objected to conditions being imposed by our employer without regard to our wishes or the real needs of agriculture. In 1927 farm workers, including nurserymen, had a five and a half day week under Orders made by the Wages Board; necessary work done outside the hours to which the minimum wages related were paid for at overtime rates. It was quite legitimate for an employee to enter into a contract of service involving a six or even seven day week provided the conditions of service did not contravene those prescribed in Wages Orders. If I had agreed to conditions of employment involving working on Saturday afternooons and Sundays then I would expect to abide by the contract. My employer and Mr. Miles had testified to my gook character and dilligence as a worker. No one had stated that I had at any time refused to work overtime when required. I left the interview convinced that what happened in 1924 had cost me the post of nursery foreman. I was extremely bitter because of what seemed to me to be an unreasonable attitude on the part of the interviewer. But it was at a time when unreasonable people could afford to adopt a rather pompous attitude to work people under their control.

This incident added to my discontent with both farm work and life in the village. In an attempt to find some way of using my leisure time I joined Sutterton church choir. I was not a communicant member of the church and was not directly asked to become one. All I sought was company and a feeling of belonging to a community. This I gained in some measure when I was invited to join a newly formed choral group. My association with these two groups was, however, brief. When the vicar in one of his sermons dealt in some detail with the definitions of a churchman I felt I had placed myself in a false position for I could not become a member of the church. So I withdrew. The village choral group gave one concert and soon afterwards lost support and discontinued its activities. I was back in the old rut and the prospects horrified me. Mother said I was the most dissatisfied member of the family and there must have been many occasions when she found me a very trying individual to have about the house.

Chapter 16.

BACK TO THE LECTURE ROOM

COMPARISONS USED, IN THE PAST, often gave the impression that working conditions in agriculture were more desirable than those in other extractive and manufacturing industries. It was argued that the close employer-employee relationship in agriculture, the open life and the varied nature of farm work gave a kind of personal satisfaction that bound farmworkers to their industry in a way not to be found in any other occupation. This was not a point of view which fitted in with my experiences of farm work and of farm workers. Most young farm workers in the period between the two world wars found themselves tied to the industry and I cannot think it gave any of them much personal pleasure. They stayed because they could not escape. Certainly I wanted to get away but was too timid, too afraid to make the effort, perhaps too ready to wait for someone to come along and offer an escape. In a way that is what happened. Early in 1927 I saw in the *Landworker*, the journal of our union, particulars of the Buxton Memorial Scholarship offered to farm workers and tenable at Ruskin College, Oxford. Knowing nothing about Ruskin College or the courses of instruction taken there I was uncertain whether to apply. The year spent at Kirton Agricultural Institute had, so it seemed, done little to advance my chances of securing a position of responsibility either inside or outside farming. I decided, however, to take a chance and applied for the scholarship. If successful it would at least take me away from farm work for a year.

Ruskin College was founded in 1899 to give working men a liberal education within a residential college. It had a special attraction for active workers in the trade union and labour movement, people likely to have strong working class prejudices and an urge to understand the economic and social problems of their day. The intention was, and I believe still is, to give students a 'genuine education as distinct from partisan propaganda of any kind' and 'to gather the windbags and prick them, letting the air escape, and filling them with substance'.*

This liberal approach to education caused some working class organisations to doubt whether they ought to continue giving financial support to the college. They were not against a liberal education for their members but feared the training would be too academic; that it would ignore the educational needs of those who wished to work within the trade union and working classs political organisations. Some critics wished students to be given more opportunity to study the Marxian analysis of the capitalist system. Others while not adverse to a wider, less partisan, training in the social sciences considered insufficient attention was given to the theory and practice of trade unions. It was argued that the non-partisan approach to economic and social problems tended to discourage students from maintaining their interest in improving wages and conditions of employment. Workers took an understandable interest in the courses taken by students in receipt of financial assistance from the T.U.C. or their own trade union. Some of these students found themselves subjected to vigorous questioning by fellow trade unionists when home for the vacations. Members of the General Council of the college as well as the staff, welcomed the critical interest taken in the college by the trade unions and other working class organisations. It was felt this would lead to a better understanding of what the college was trying to do and gain from the trade union greater financial support. It was also felt it would result in trade unionists urging their local authorities to provide more funds either by way of grants to the college or by sending more students there.

As with the Ministry of Agricultural Scholarships a grammar school education was not essential. The selection for the Buxton Memorial Scholarship was by competitive examination. This almost put me off applying. For convenience candidates were asked, individually, to name a local person of standing willing to

---
* *The Story of Ruskin College.* Oxford University Press. Revised Edition 1968, pp. 4-9

supervise the examination. This was done to enable candidates to take the examination near their homes and avoid the expense of taking it in Oxford. I named our local stationmaster and was later informed that the examination would be taken at his home.

Some of the questions set called for a greater knowledge of economic and social history than I had. I remember that one asked candidates to state what they understood by the term industrial revolution. In my mind revolutions were bloody affairs and I was not sure whether the industrial revolution was a particular historical event similar in character to the Russian revolution. Later when I learned something about the economic and social changes covered by the term industrial revolution I was glad I had not attempted that question. The paper was set and marked by Professor A. W. Ashby of the Department of Agricultural Economics at the University College of Wales, Aberystwyth. He had been a student at Ruskin College for two sessions during 1909-1912. After leaving elementary school he had worked for some time on his father's small farm in Warwickshire. His formal education may not have been much better than my own but his home background had given him wider opportunities for continuing his education after leaving school. His father, keenly interested in the economic and social problems of small farmers and the countryside, was an active supporter of the Liberal Party and a frequent contributor to local newspapers, magazines and journals including the Economic Journal.

Although Ruskin College was not a constituent college of the University of Oxford it had close association with the other colleges and its students were allowed to take the university examinations for the Diploma in Economics and Political Science. This Ashby did at the end of his second year. After passing these he was, on the recommendations of his tutors and some university dons, appointed a tutor for the Workers' Education Association. Their experience of the W.E.A. and university extension classes persuaded them that Ashby was the kind of tutor required in the field of adult education. His home background, his balanced approach to the study of economics and social problems as well as his easy manner in argument marked him out as one able to command the attention and support of working class students with minds conditioned by strong working class prejudices on matters relating to the causes of, and remedies for, their own discontents.

In 1913 he gained a research scholarship, the first to be offered by the Board of Agriculture. This enabled him to spend one year at the Agricultural Economics Research Institute, established in Oxford in 1913. From there he went to U.S.A. where he spent a year at the University of Wisconsin. After his return from America he joined the staff of the Agricultural Economics Research Institute and spent some time, after 1917, working on problems concerned with the control of cereal prices and farm wages. In 1924 The Ministry of Agriculture, acting on the recommendations of the Departmental Committee on the Distribution and Prices of Agricultural Produce, made funds available for the establishment of agricultural economics advisory centres in different parts of the country. Ashby was appointed head of the Welsh centre at Aberystwyth. Three years later the University of Wales conferred on him the status of Professor, thus creating the first Chair of Agricultural Economics in Britain.

Few other people who might have been invited to undertake the task of seeing and marking the papers for the Buxton Memorial Scholarship had a better appreciation of the difficulties facing young farm workers taking the examination. Some might have sought to test candidates' knowledge of school subjects but Professor Ashby was more interested to discover whether they had any knowledge of the economic and social problems of their rural areas; whether they had taken any interest in the historical changes that had taken place and in their effect on agriculture in their own district. He wished to know what candidates had made of the opportunities available to them to improve their education. Some of the questions invited examinees to give an account of the action taken to keep themselves informed on current economic and social developments. Others asked them to name some of the rural voluntary organisations and give their views on the contributions these had made to the improvement in social life in their village.

The result of the examination was announced in the local and national newspapers under the heading **From Plough to College**. All reported Professor Ashby's comment that 'the first eight papers were a tribute to the class of farm worker and incidentally to the elementary schools'. I think he was over generous to the candidates as well as to rural elementary schools. I was naturally pleased to be the successful candidate but this pleasure was tempered by a feeling that many young farm workers could, had they taken the examination, bettered my performance. So little had been done to encourage young farm workers to continue their education after leaving elementary school.

No evening classes were provided in our villages and I think that must have been so in many other areas. Only those who lived within cycling distance of a town could take advantage of the limited facilities for continuing their education. The absence of these facilities bred in young people a belief that they lacked the qualities essential to benefit from attending evening classes. The need was to convince young people that they had the capacity and given the facilities could improve their economic and social position. Unless this could be achieved there was no hope of anyone obliging their local education committees to provide evening classes within easy reach of young people living on farms. Unless people fully understood the value of education to themselves and their children there was little hope of grants being provided to enable boys and girls to go to grammar schools, to technical institutes and colleges.

The boost which gaining the scholarship gave to my vanity was soon lost when I came in contact with my fellow students at Ruskin College. I think the college had a total of 30 students; the small number in those years was in large measure due to the financial difficulties experienced by trade unions and local education committees. Few trade unions had the resources to send promising young workers to Ruskin College for a year and only one or two local authorities gave loans or other financial assistance to those wishing to take advantage of the training provided there.

Most students hoped to be able to stay for two years either to study for the University Diploma in Economics and Political Science or for qualifications which they hoped would enable them to take positions as social workers or probation officers. A small number went there for one year's intensive study of a particular subject. No pressure was placed on students to take the Diploma examinations but tutors favoured the practice as one way of encouraging them to take their studies seriously.

Before I went to Ruskin I had met no one with a direct knowledge of the courses of study provided, no knowledge of its close association with trade unions or with the University of Oxford. This may have been because I had, after going to Kirton Agricultural Institute, been less active in Boston's Trades Council. I don't remember any mention of the college at the monthly meetings of my branch of our union. Having gone there ignorant of the conflicting views on its functions as a residential centre for adult education I was taken aback when I met Mrs. Uzell, a member of the executive of our union and heard her criticism of the college and its students. In her view students ought, after the completion of their studies, to return to their old occupations and use their training to further the work of their own and other trade unions. She complained that too many students, including those financed by trade unions and the T.U.C., left Ruskin and lost all interest in the struggle of their old workmates. This was much too sweeping. A few students did enter a university. But this did not necessarily mean they lost interest in working class organisations. In the inter-war years the majority of students, after leaving Ruskin secured appointments either with their trade unions, with one or other of the Constituency Labour Parties or with organisations and institutions concerned with problems affecting the wellbeing of working class families. One had to recognise that not all people are so dedicated to the task of improving the lot of their fellow men as to be prepared completely to ignore opportunitiess for personal advancement. Some, who took full advantage of the opportunity to continue their education after leaving Ruskin, and subsequently secured professional appointments, showed they could, as a consequence, give greater service to the trade union and labour movement. George Woodcock was an outstanding example. After spending his academic years at Ruskin College and then going to Oxford University he was for many years Géneral Secretary of the T.U.C.

Since the University of Oxford and some of the dons assisted Ruskin College in a variety of ways trade unionists feared Ruskin would lose its identity as a college catering for working class men and women having little hope of gaining entrance to a university. There was strong objection to what was regarded as undue emphasis on courses which enable students to take the University Diploma in Economics and Political Science. Critics considered scholarships given by trade unions and the T.U.C. should be granted for only one year. This would give more workers a chance to train for work in their unions. It was considered that by limiting residence to one year the college would pay greater attention to subjects more directly related to the needs of active trade unionists and of workers in the labour movement, in particular to training in public speaking, the work of trade union organisers, shop stewards and industrial negotiators. Such a shift of emphasis would have meant abandoning the ideas and hopes of the founders of the college. In the event of such a change no one could be sure that the Board of Education would have continued its financial grants to the college.

When the college first opened in 1899 students had to do all the domestic chores including the preparation of meals and scrubbing floors. Long before my time the duties had been reduced to serving meals, washing up afterwards, and keeping one's room tidy. A porter and kitchen staff attended to all the other domestic duties. Matron made daily rounds of rooms to see that beds had been made and that the rooms were tidy. Teams of three students took in turn a week's duty of serving meals and washing up afterwards. This was a new experience for many of the men and for some of the girls. I don't think anyone liked the tasks but accepted them as part of the traditional life of a Ruskin student. The odd one sought to avoid having to wash dishes. I remember one in the same team as myself who objected to putting her hands in hot washing-up water. She managed to bribe the other male member of the team and myself to wash the dishes. With so many greasy dishes we used a lot of soda and, having no rubber gloves, this was hard on the hands. But those of us who had spent some years in manual work, not exactly kind to hands, never worried about the damage soda might do to them.

This was my first experience of institutional cooking and feeding. I had no complaints, in fact I enjoyed the wider variety of food. At home there had been little variety so I was glad to be where I had more hot meals, morning, noon, and evening. Except for two students who were vegetarians and whose choice of food seemed to me rather limited, the rest of us were well served. There was little choice for individual meals, we had to take or leave what was on offer at any particular time. As between one day and another there was plenty of variety, I had never been allowed to waste food and was not finicky. I made a point of studying the likes and dislikes of students and when the main course at lunch or high tea was to my particular liking I always hoped to sit next to a fellow student who disliked the course. If successful, I managed to get a second helping. I was never hungry, never felt the need, as some others did, to go into the city for cups of tea and cakes. We had student meetings from time to time at which complaints were made about food. Listening to these I often wondered how well some had fed during their long periods of unemployment. But it is common for young people brought up on bread and jam or fish and chips to experience some initial, if not permanent, difficulty in accepting a more varied and balanced diet.

Chapter 17.

AFTER EARLY DOUBTS — A DIPLOMA

AT FIRST I EXPERIENCED the same kind of difficulties as those which had troubled me at Kirton Agricultural Institute. I could not concentrate on reading for any length of time. This worried me since I found I had to do much more reading than when at Kirton. However, the subjects held my interest. The social sciences, history, economics and politics, directed my attention more directly to human activities and problems. I was interested in people rather than in plants and farm animals. I also found the language of the social sciences, at least the words if not at times the meanings attached to some of them, more familiar than those of the physical and biological sciences. It was also an advantage to be living with other students. Being in a small community of people pursuing the same subjects enabled me to join in discussions on matters dealt with in lectures. I received a great deal of help from my fellow students and from tutors.

Students came to Ruskin from varied home and occupational backgrounds, with varying experiences, and with differing levels of formal and informal education. Most of them came with practical experience of some particular industry or branch of commerce. Our tutors, though lacking any detailed knowledge of our past education, knew from their long association with working class students that some of us would have serious defects of language. Consequently all new entrants had to take a simple test. This was in parts amusing: for example we were asked to state what was wrong with two statements, one, 'The hand that rocked the cradle has kicked the bucket', the other, 'Lor' bless yer Gov' I turned as white as a bloody sheet'. It was easy to see the faults but in explaining them I fear I must have exhibited, in an extreme form, my own incompetence to write good English in a pleasing style.

I found it extremely difficult to rid myself of ungrammatical colloquial English learned as a child at home and on farms. The language of the social sciences with its jargon and specific definitions attached to some ordinary words such as rent, interest, wages and profits, was a new experience. In the early weeks I was often in trouble because the meanings which I attached to some words differed, either from those given in the dictionary, or from those of my tutors and students with previous knowledge of the subjects. Having to write one and on occasions two essays a week obliged me to do a great deal of reading and to waste a great deal of paper as I struggled to express myself. In earlier years I had never done sufficient wide reading to acquire a satisfactory vocabulary for writing correct and acceptable English. Often I could not find the right word in my own mind or in the dictionary to express my views.

Each student was assigned to a tutor at the beginning of each term. First year students with serious language problems had Mr. Schofield as their tutor. I was assigned to him and I found it a devastating experience, not because Mr. Schofield was unkind or excessively harsh in his criticism of my work but simply as a result of becoming aware, for the first time, of my deficiencies. At each weekly tutorial, which lasted an hour, I read my essay to Mr. Schofield and at the end left it with him in order that he might make some attempt at correcting some of the more serious mistakes. For some weeks these essays came back to me badly mauled in red ink. George Woodcock, former General Secretary of the T.U.C., has given an account of his experiences with Mr Schofield in 1929 when he was a student at Ruskin. Referring to the first essay he submitted he wrote: 'The essay flowed easily from my pen and . . . I fully expected it to establish me in the first flight of new students. Schofield quietly but mercilessly tore it to shreds, but from then on my essays, though harder to write, were less offensive to read!' I cannot think George Woodcock had his first essay so heavily splattered with red ink as all mine were during my first term in Ruskin. In my case subject matter, grammar and style met with severe criticism. On occasions I left Mr Schofield at the end of a tutorial almost reduced to tears. Once

I was so badly upset that Mr Schofield came in the afternoon and took me out into the country in his car, a chain driven Trojan. During the drive he gave me good advice and encouragement. Essays presented to other tutors later in the courses also met with a great deal of criticism.

In addition to the weekly essay for my tutor I had to write others at intervals for one or other of the weekly seminars. Reading these essays to fellow students was in some ways more devastating to my morale than the weekly encounter with my tutor. He, with his knowledge of his subject and long experience of working class students, had a way of presenting criticism which I could take without too much upset. The language used by tutors in their criticism was temperate. Students themselves showed less consideration: their criticism, though given without any deliberate intention to be malicious or offensive, was on occasions presented in a fashion which appeared to be excessively ill-mannered. Limitations in the choice of words often made the language of my fellow students seem brusque and doctrinaire. To a large extent they brought into the seminars the vigorous, uncompromising, and often uncomplimentary language used at trade union meetings. Bringing together working people from different parts of the country, from different industries and occupations had its problems. The brusque speech of the northerner was upsetting at first but one soon discovered it came from kindly people. I managed to get through the first term without serious nervous upset. I had gone to Ruskin well nourished and in good health. One or two other students from the depressed areas did not fare so well. They were physically ill-equipped for the strenuous mental tasks and during that term had nasty attacks of nerves. At one stage one lad was so depressed that matron decided he ought not to spend his nights alone. I spent a couple with him by which time he had pulled himself together.

A number of students from the Potteries and mining areas had, previous to coming to Ruskin, attended W.E.A. and University Extension Classes in subjects such as literature, economics, philosophy, psychology and political theory. They had gained a good grounding in some of the subjects taken at Ruskin. Others, like myself, without this previous training, had to work very hard and had to rid our minds of many crude notions previously held on matters relating to the economy and government of this country. I could not hope, in the early weeks of that first term, to produce work of the standard achieved by those who had taken courses in some of the subjects before coming to Ruskin. My fear was that as the courses progressed the gap between my performance and that of other students would widen. It was difficult to accept the advice of tutors not to worry too much about my deficiencies. They feared that students with no previous formal training in the subjects being taught might spend too much time studying and neglect to take sufficient physical exercise and rest. There was no risk that I would burn too much of the 'midnight oil'. Most evenings I found it impossible to study after 10.00 p.m.

From its earliest days Ruskin College had followed the university practice of taking lectures, tutorials and seminars in the mornings and early evenings, leaving the afternoons free for recreation. We had no facilities for outside games and spent our afternoons either walking in the country or browsing in Oxford's bookshops. The walks gave me particular pleasure for, despite my rural upbringing, it was a new experience to be able to get away from country roads, to cross fields and walk in woods. Our countryside is flat and open, in some districts hedgerows have been removed and ditches filled in to increase the size and improve the layout of fields. We see plenty of sky, many church spires, but have to go miles into the Wolds or to a neighbouring county to find positions which give panoramic views of fields and woodlands. I liked the afternoon walks around Oxford.

I went home at the end of the first term in a doubtful mood about my future. My friends would say that I was always a 'doubting Thomas'. I felt I had made little progress, apart from discovering a few of my deficiencies including my inability to express myself in an acceptable language. During the Christmas vacation I spent one evening with our local station-master and his wife. As I mentioned earlier he had supervised the scholarship examination. I explained to them my difficulties and my feeling that I ought not to return to college after the holidays. Later in the evening when Mr. Massey went out to attend to some business his wife gave me a thorough dressing down, charging me with lack of courage and determination. In her view disadvantages were a test of character, obstacles which one had to seek to overcome. Not to make the attempt was a serious fault in personality. At the end of this severe admonition I felt ashamed of myself but was not convinced that I could succeed. After a while Mrs. Massey described a person to me and asked if I knew him. Her description reminded me of a gentleman who used to visit the farm where we lived when I was a very small boy. My mental picture was of a man wearing a longish jacket and a hard hat green with age. To me, at the time, he seemed an old man but to a lad of five anyone over 50 would be old. Mrs. Massey did not say she had met the person she described and no names were mentioned. If she had met the man I had in mind she could not have known that he had met me. She told me he had given her a message for me, which was to the effect that I would succeed in what I was attempting. I don't presume to know anything about the possibility of communications between the living and those who have passed on, I merely state what passed between us on that evening. Knowing Mrs Massey as I did I am sure she did not make up the story for my benefit. I went back to Ruskin after the holidays, the shame of giving up after that experience would have been greater than any failure in the examinations.

With the consent of the university and colleges Ruskin students attended courses of lectures given in the Examination Schools and in some of the colleges. Few of our first year students attended these lectures, their time being fully occupied by foundation courses in English, general history and economics. Those with some previous experiences of W.E.A. and university extension lectures did attend one or two terminal courses given in one or other of the colleges. In my second year I attended two full courses given in the Schools. Unlike undergraduates we, and students from the Workers' Catholic College, did not wear academic dress. We had, when attending the first lecture of each course, to present a note to the attendants which gained us admittance.

Our tutors advised us to attend a course of lectures on economic theory given by Mr. G. D. H. Cole, another course on industrial history given by Mr. E. Lipson and a third on systems of representative government given by Professor W. S. Adams, warden of All Souls. I was not alone in failing to attend the full course given by Mr. Cole. His delivery was too fast and my understanding of economic theory too elementary. I found myself constantly losing contact with the argument. Mr Cole's reputation with students was high and he attracted large audiences. At the beginning of his course students filled the West Room and overflowed into an annex. One had to be there early to be sure of a seat near the dais. At the beginning Mr. Cole advised us not to take notes but instead to concentrate on his presentation of the subject. So far as I was concerned the advice was unnecessary, he spoke so fast that the few notes I did take proved useless. After about three lectures he again advised students to close their notebooks and concentrate their full attention on the lectures. If, he said, we did not understand any part of his arguments we should ask for further explanation. He suggested the number of students attending the lectures was, at that stage, too large for useful interruptions but he seemed confident attendances would decline to a level at which questions would be welcomed. I don't know how many students completed the full course. I gave up very early, having to admit that his treatment was too advanced, too abstruse for my sluggish mind. I decided my time would be better spent reading Henry Clay's *Economics for the General Reader*.

While lecturing Mr. Cole walked to and fro on the dais. This I found distracting. I wondered whether he would take the final step and fall off. However, while I attended his lectures neither he nor I came 'down to earth'. He never referred to notes, he was just a talking machine from which words shot out like bullets from a rifle. In contrast I found his seminars on industrial organisation most helpful. He invited a mixed group of undergraduates and students from Ruskin and the Catholic Workers' College to attend these seminars. We felt it a great privilege to be allowed to do so. Mr. Cole guided us through a study of the basic problems facing the heavy industries at that time. His masterly way of presenting conclusions based on masses of published data in a concise and understandable manner was most valuable to students who, like myself, had opted to take modern economic organisation as one of their subjects in the Diploma examinations as well as to undergraduates reading for the degree in 'Modern Greats'. At each seminar Mr. Cole introduced the facts and problems of a particular industry before inviting contributions from the students. Ruskin students, especially those who had worked in the industries being examined, were usually the first to contribute in the open discussions, often with some criticism of owners, managers and governments. It takes a little time, even in the intellectual atmosphere of Oxford, to rid oneself of ingrained prejudices; to train oneself to examine all the evidence before giving a final verdict. We learned from these seminars the varied and complex interaction of a wide range of circumstances; that few economic and social problems are adequately exposed for an analysis to lead to the formulation of policies aimed at improving the efficiency and social conditions of industry.

Ruskin students had in their various occupations, and as active members of their trade union and constituency labour parties, become accustomed to seeing causes of, and remedies for, economic problems from the narrow view of their own class. Few took a wider more detached view. Most of us had to train ourselves to stand apart from our own particular group problems in order to appreciate fully the national need.

Mr. Lipson, economic historian, had his peculiarities. The time given to each lecture never varied. We knew the exact time at which he would enter and leave the West Room. Each lecture lasted for fifty minutes. I had the impression that he read extracts from either extensive notes or from manuscripts of books and articles. Being rather short of stature he seemed to stand on his toes as he delivered his lectures. His delivery was slow and clear and I had no difficulty in taking notes. He never seemed to get emotionally involved with his subject. I remember one instance of this when he was dealing with the social consequences of technical changes in the eighteenth and nineteenth centuries; changes which caused many people to lose their economic independance and for some their livelihood. Mr. Lipson told us in a monotone: 'and these changes had a deleterious effect on the workers'. This statement and the manner of its delivery seemed to those who had experienced long periods of unemployment unduly detached. Perhaps it was important for students from working class homes, paticularly those from depressed areas to have the facts of social history presented dispassionately. It has never been easy for men whose livelihood is endangered by technical changes to get any comfort from being told that particular changes are in the long term interests of workers as well as the nation generally. 'In the long run' was a phrase which, when used by economists, hit Ruskin students like a cold blast of air from a fireless chimney on a cold winter's night.

Professor Adams gave his lectures in the East Room of the Schools. He came into the room, mounted the dais and sat for a few moments before starting his lecture. This was a practice which also never varied during the course of lectures. He barely looked at his notes before pushing the chair and table to one side; then, standing at the rail, away from his notes, he lectured quietly, in a pleasant voice without hesitation, without any fumbling for words. One sensed the lecture had been given many times before. He held my interest even when I was not wholly convinced that a hereditary head of state was preferable to one elected by universal franchise. He left us in no doubt of his preference for the British system of government. He contrasted the flexibility of the British system with that of some other countries with written constitutions. He impressed on our minds the old cliché that the British constitution bends to the will of the people and does not break. This knowledge gave us little satisfaction for we wished to learn how to minimise the influence which private wealth and property had on elections; to know how to eliminate the undue authority which some employers exercised over the political freedom of their workers.

We had been advised by our tutors to attend other shorter courses of lectures in the University. One which some of us attended was given by Miss Butler in Barnet House. I cannot remember the exact title of this course but it was concerned with industrial organisation. As an adjunct to the course Miss Butler arranged visits to industries and institutions. I remember we visited Chipping Norton Workhouse, Littlemoor Mental Hospital, Whitney Blanket Factory and a farm near Banbury. At the workhouse we saw the accommodation for casuals, as well as the provision made for other inmates. With so many men of all ages walking the roads in search of work the casual ward was fully occupied each night. None could stay for more than one night and each had to do the required stint of work before he left. Some did the work on the day of arrival, this allowed them to get away earlier in the morning hoping to improve their chances of finding work on farms. Others avoided work as long as they could. Understandably those who had lost all hope of finding work drifted from one workhouse to another arriving late in the evenings and, if possible, slipping away in the mornings before doing their allotted task of work. We were interested in the food and accommodation provided for casuals. The evening meal consisted of tea, bread, margarine, and cheese, It was the bare minimum. Ruskin students in the party concluded from the visit that nothing in the provisions could give these unfortunate men the feeling of having the sympathy of the general public, We felt they were looked on as scroungers and treated as such.

At the mental hospital the doctor in charge gave us a well deserved lecture, prompted by hearing a student refer to the hospital as a lunatic asylum. The doctor reminded us that his hospital and its patients, like other hospitals and their patients, needed our sympathy and help not ridicule. At that time many people thought of mental hospitals as places where those suffering from mental illness could be put away and perhaps forgotten. Some people felt ashamed when a member of their family had to be 'put away'. We were directed by the doctor to erase from our minds the old view of this kind of hospital otherwise it would never be possible to return any of the patients to their homes. Many inmates would, with sympathy and care from relatives, regain full confidence in themselves and again live useful lives within the community. Even those with permanent disabilities were people who should not be hidden away from the rest of society for the remainder of their lives. We left the hospital chastened but better informed on the work being done at Littlemoor Hospital to help patients and made more aware of our individual and collective responsibilities for the care of these people.

Having no experience of the textile industry I welcomed the opportunity to visit Whitney Blanket Factory and see at first hand the processes involved in the manufacture of blankets. After visiting this factory I had a better understanding on the points made by Mr. Lipson when dealing with the technical changes which took place in the woollen industry during the eighteenth and nineteenth centuries. The visit to a farm in North Oxfordshire also proved to be of particular value to me. From it I learned that after working for nine years on farms and after spending a year at a farm institute there was still a very great deal which I did not know about the practical problems of British agriculture. The visit underlined again the disadvantages of spending a year at Kirton Agricultural Institute rather than at a farm institute in another part of the country where soil conditions and farming systems differed markedly from those of South Lincolnshire. As we walked the farm I made a comment not exactly favourable to its management. The farmer must have heard me for he asked if I knew anything about farming. I told him I had worked on farms in Lincolnshire. He wanted to know which part of the county and when I gave him the information he told me that any fool could farm there. He was right to take me to task for passing judgements without any knowledge of the problems of farming on the Oxford clays.

At the end of my second year in Ruskin I, along with other students, sat the examinations for the University Diploma in Economics and Political Science. These involved six three-hour papers taken on three consecutive days. Undergraduates taking them had to wear full academic dress. Students from other educational institutions outside the University who had been permitted to take the examinations were required to wear white bow ties. That was as far as we got to wearing academic dress. I don't know how well I did in the examinations. One student from Ruskin secured a pass with Honours and went into the University where, after a further two years, he secured First Class Honours in Modern Greats. I was content to know that I had not failed the examinations.

Chapter 18.

IN HARMONY WITH THE UNIVERSITY

IN THE EARLY DAYS of Ruskin College the relationship between its students and university undergraduates had been anything but harmonious but during my days there it was good. Various undergraduate societies invited us to their meetings and there was always an eager demand for the two invitations which Ruskin received to attend Union Debates. Some Ruskin students found it too irksome listening to debates when they wanted to take part in the battle of words and for this reason rarely made requests for tickets. Each of the main political parties had their university association. Ruskin students were, almost to a man, members of The University Labour Club. Our number was too small and largely unknown to the general body of undergraduate members to gain positions as officials of the club. Perhaps also the critical attitude of the most vociferous members from Ruskin persuaded other members of the club to seek their officials from undergraduates who had less extreme views on the capitalist system and the priviledged class. There was fairly strong support from Ruskin students for the view that discussions on economic and social problems were too academic. Men and women who, prior to going to Oxford, had experienced long periods of unemployment or who had first hand knowledge of conditions in the depressed areas, showed little patience with theoretical and philosophical arguments on matters which in their view had little relevance to the urgent human needs of working class families in the 1920s. For these reasons some of my fellow students used every opportunity to direct the attention of members of the Labour Club to problems of immediate concern to their own people. This, at times, caused a great deal of heated argument between the two groups; undergraduate members frequently advising Ruskin students to get off the slag heaps of the coalfields and take a wider, less personal view of the problems facing Britain and the rest of the world. This was asking a great deal of people whose parents and other close relatives, as well as old workmates, had been deprived of the right to work and maintain themselves as self-respecting human beings. It was not easy for them to enjoy the comfort and comparatively good living at Ruskin knowing that at home their relatives had to accept idleness and general deterioration in health.

The economic and social problems of unemployment during the 1920s occupied an important place in discussion by many undergraduate societies and many young people had come to the conclusion that we had more people in Britain than could be given useful full time employment. The growth of population and size of families were two particular questions examined in some detail by one society, I think it was the THOMAS MORE SOCIETY. It presented its conclusions at a meeting to which students from Ruskin as well as representatives from some other organisations had been invited. The report recommended that the total population ought to be stabilised at around its then number. In order to achieve this it was suggested that advice on birth control ought to be provided by local authorities, especially to the poorer sections of the community. There was nothing revolutionary in these recommendations. I mention the incident to illustrate the interest taken by students in important social issues of their day.

During the late 1920s and early 1930s unemployed workers organised demonstrations and marches to draw public attention to their plight. Many people considered that marches did not, and could not, help the unemployed. Many, fortunate to be in work, had become so concerned to hold on to their jobs that they gave little effective support to the demands of those out of work. The situation had, however, become so desperate that marching on London seemed, to the unemployed, the only way of arousing the consciences of the general public. They hoped it would draw attention to society's failure to ensure that every able bodied person should contribute to, and share equitably in, the wealth produced. It was their only means of demonstrating against man's inhumanity to man. One march to London took place during my time in Oxford and a large contingent of unemployed from South Wales spent a night in Oxford Town Hall. Ruskin students along with other people from the University and City assisted in the work of feeding and caring for these men during their overnight stay. The march imposed additional hardship on men already suffering from severe undernourishment. I had little experience of being out of work and was completely ignorant of the damage done to human beings by long periods existing on the very low rates of unemployment benefits and, after exhausting these, having to submit to a means test when seeking extended benefits. Talking to these men one felt ashamed of allowing people to suffer the indignity of enforced idleness without any hope of becoming involved in the production of the essentials of the good life. These men had a right to be bitter. We had obliged them to become wholly dependant on the miserable weekly handouts from the Ministry of Labour. Some people too readily accepted the view that many of the unemployed were shirkers. I was surprised that none of the men on that march showed any bitterness. Many of their leaders were communists but I never heard any expressions of the kind one might justly have expected from them.

This experience made me realise my own good fortune in escaping such hardships and in having friends who gave help and encouragement when I was depressed about my future. I must mention two in particular to whom I have always been grateful. I have already referred to Professor Ashby and will mention him again. Meantime I must refer to T.W. Price. As a Rochdale millhand he became a member of the first tutorial class organised by the Workers' Educational Association in 1907. The tutor of this class was economic and social historian R.H. Tawney. From him Mr. Price not only improved his knowledge but also acquired some of the important skills of lecturing to working class students. A few years after attending these classes he became District Secretary of the W.E.A. for the Birmingham Area. Then in 1919 he was appointed the first Warden of Holybrook House, Reading, which had been established as a W.E.A. centre for training selected students from W.E.A. and University Extension Classes to become tutors of one year classes and leaders of study circles. He was a fortunate choice as a W.E.A. tutor and as Warden of Holybrook House. In addition to his practical experience of industry he gained tremendous pleasure from his contacts with adult students determined to improve their knowledge of the arts and sciences, of philosophy and religion. Most lecturers find adult students more stimulating than younger ones with little or no practical experience of earning their living. The wide range of practical experience and information which mature students brought to the classes was an asset welcomed by lecturers ready to use it as an aid to their lectures. Each student had his own particular occupational experiences, his own well established prejudices. It was more difficult, as well as often more rewarding, to lecture to adults because their sole purpose in attending classes was to gain knowledge, to satisfy their curiosity. They attended week after week, year after year only so long as lecturers demonstrated a deep knowledge of their subjects and the skill to present this in an interesting and provocative manner. University students with little knowledge of conditions outside classrooms and home often failed, through no fault of their own, to give the same mental stimulation and discipline as that given by mature students. Lecturers of W.E.A. and University Extension Classes could expect, and if wise welcomed, interruptions during lectures. Not all interruptions were relevant to the particular matter under examination but being prompted by something a lecturer had said were accepted as part of the exchange of ideas. An experienced lecturer was never surprised or upset by the odd individual who joined a class merely for the opportunity to air his own views on almost every subject. Some students had strongly held opinions on a wide variety of subjects and a well seasoned tutor used these over assertive, overconfident students, to provoke more reticent members of his class to add their contributions to the discussion.

Personal qualities of lecturers are important and some practical knowledge of industry or commerce is an added advantage. Mr. Price had the right personality and practical experience of the textile industry. He knew many adults tended to think through their bellies rather than their heads but this was to be expected of people whose legitimate family needs often exceeded their incomes. He had the natural gift of friendliness which enabled him to use the experiences and prejudices of members of his classes without causing resentment or embarrassment and the ability and authority to win the confidence of slick debaters as well as of serious students. The hour's discussion which followed each lecture was a testing time for lecturers. Mr. Price demonstrated during these discussions his skill in holding the balance between conflicting opinions of members of adult classes. In the 'what is' and 'what ought to be' kind of arguments on economic and political matters there is always plenty of scope for differing opinions. When some members of a class have rigid, inflexible minds the skill of a lecturer can be put to a severe test. Mr. Price exercised the kind of control which satisfied members that they had, at the end of an hour's discussion, had a fair innings and had gained a better understanding on the views of other members of the class. As Warden of Holybrook House he organised two summer schools each year, one in July and the other in August. The official number of students attending each school was limited to twelve, a number which Mr. Price argued allowed differences of views to play a useful part in the work of the school without causing the formation of cliques. He was well aware of the tendency within working class organisations for differing opinions to result in the build-up of turbulent and at times destructive group action. Each student received free board, lodging and tuition. In addition to Mr. Price two other tutors came for the period of the schools.

It was unusual for a student to attend more than one of these summer schools but on occasions exceptions were made. No one was prohibited from applying for permission to attend a second or third school. If the number of suitable students was greater than the available places then anyone who had already been to one of the schools was unlikely to be offered the further opportunity of a month at Holybrook House. The odd candidate who it was considered had particular qualities or needs, might, if circumstances allowed, attend more than one school. I could not claim to have any qualities which marked me out as one likely to be of particular value to do the work of the W.E.A. but I certainly needed extra help with my studies. This may have been the reason why, during each of the summers of 1928 and 1929, I attended two summer schools. During my stay there Mr. Price gave me a lot of his time; perhaps he considered I was in special need of his help.

Throughout the four weeks of each school student had to prepare an outline of twenty-four lectures on a subject of their choice. One of these lectures had to be prepared and delivered to the class. After the lecture one had to face criticism from tutors and students on the prepared syllabus and delivery of the lecture. Students who had attended a number of W.E.A. and University Extension Classes tended to be severe critics of beginners' attempts to master the art of lecturing. Tutors were less devastating in their criticism. They avoided making comparisons between the performance of students and experienced lecturers. As I had attended very few lectures and had little experience of public speaking I did not find the task easy. Arrangements were made for some afternoon visits to local industries and to places of educational or historic interest. One of the best afternoon outings was a trip which each school made to Goring Woods. We took the train from Reading to Pangbourne and them went by punts up the Thames to the woods for a picnic tea. The beautiful countryside around Reading gave me great pleasure perhaps because it was in sharp contrast to the flat open country of South Lincolnshire.

Mrs Price had full responsibility for the domestic side of the schools. Students made their own beds and kept cubicles or bedrooms tidy. They knew the minimal rules of the house had to be observed. No one was allowed to take meals unless properly dressed, wearing jackets, collars, and ties. The occasional student who had to be reminded of the rules may have thought Mrs Price rather too particular, too anxious to observe middle class conventions, but there was never any revolt. She was a kindly person and managed to get her own way even when young people considered some of her 'musts' to be outdated oddities. I think students loved her because she behaved as good mothers did towards their grown up children. She applied the same discipline to her three sons: the 'home' part of the school was her domain and within it her 'children' conformed to her wishes.

Over the years I have had good reason to be thankful to Mr and Mrs Price for the help they gave me during those two summers. At Holybrook House I had access to all the books I required and apart from the lectures we had adequate facilities for private study. After my stay there in the summer of 1928 I returned to Ruskin College for my second year with greater confidence in myself. I felt I had made progress and was persuaded that given fair luck I stood a reasonable chance of coping with my studies. I still had no clear idea what I wished to do after the completion of my second year. One heard of the ambitions of fellow students and of the kind of work that past students had obtained after completing their studies. Some became trade union officials, political agents, probation officers, welfare workers, W.E.A. lecturers and a few returned to their former employment. A small number became M.Ps. I could not see myself as a trade union official nor political agent. I had done no public speaking before going to Ruskin and very little while there. I never felt I had the essential qualities demanded of politicians. Public speaking on narrow political lines never came easy to me and I lacked the ego, the self assurance, necessary to convince listeners that one believed, absolutely, in the rightness of the policies one was advocating. My vocabulary was too limited, I had not, at the tip of my tongue, a wealth of denunciatory words which make a speaker, if not his audience, feel good.

The life of a politician seemed too precarious and I was not resilient enough to face the prospect of fighting endless elections, perhaps never succeeding and, if successful, of having to live with the risk that at the next election I would be looking for another job. The life of a political agent seemed equally risky and excessively demanding. One had little hope of being appointed agent to a safe

Labour constituency unless one had either an outstanding reputation as a speaker and organiser or had many influential friends in one or other of the constituency Labour parties. Applicants with little experience of public speaking, if successful, found themselves appointed to constituencies with few fully paid up party members and with substantial debts: or to constituencies which rarely had a prospective parliamentary candidate except for a few weeks before each election.

I had seen too much of the problems of the labour agent in my own constituency. During the years when Mr W. S. Royce was our M.P. the local party had no financial worries but after his death we had to depend to a considerable extent on the financial support from individual members and from local branches of the trade unions. Agents had to spend so much time on fund raising that they failed to give adequate attention to the really important business of maintaining a strong individual membership of the party. This neglect of administration and organisation resulted in further loss of members and of financial support from the trade unions. There seemed to be no escape from the vicious circle of neglect and declining membership. The Party was reduced to a small number of faithful members in villages and towns who spent their time thinking up new gimmicks for getting money to pay the agent's salary. At one stage the agent and a small group of members used all their free time selling packets of tea for a small fianancial gain and a good deal of criticism from managers of local retail Co-ops who argued that trade was being taken from them. Eventually the agent's appointment had to be terminated because ¹/4 lb packets of tea failed to encourage tea drinkers to join the Party or purchase the tea in sufficient quantities to yield the funds needed to pay his salary and expenses. I did not relish the possibility of being in that situation, or of being subjected to constant criticism from members who preferred to stay at home, near a cosy fire, rather than attend meetings or walk streets knocking on doors canvassing for new members. There was always plenty of criticism from people who supported the Party at an election but saw no reason why they should pay for their politics. These critics, after an election, told us that if we had worked harder, and had done more canvassing, the constituency would have been won by our candidate.

If I had at any time had hopes of securing an official position in my union I soon put that aside after paying a few visits to Mrs Uzell. I greatly admired this lady for she had given so much of her time fighting on behalf of farm workers and their families. She had every right to expect those fortunate enough to get an opportunity to improve their education to assist in the work of trade unions. But I feared that if I went back to farm work it would be the end of any hope of escaping from that life.

Chapter 19.

TOWARDS A UNIVERSITY DEGREE

FROM INFORMATION GAINED during my first year at Ruskin I knew it was possible, if over the age of twenty-three, to be admitted to some universities under special regulations and allowed to take a degree. In the first term of that year I met Mr W. J. B. Hopkinson, a former holder of the Buxton Memorial Scholarship. After gaining the Oxford University Diploma in Economics and Political Science he had been awarded one of the Ministry of Agriculture senior scholarships and on the basis of his diploma was accepted as a student by the University College of Wales, (U.C.W.) Aberystwyth. He was allowed to take courses for the B.A. degree in Economics with Agricultural Economics and at the end gained a first class Honours degree. He gave me information on the courses he had taken and also details of courses taken by other adult students at Aberystwyth who had entered the College without having taken the normal matriculation examinations. I also gained from him a general idea of the kind of work undertaken by agricultural economists and learned more about their work. In my second year Mr C. S. Orwin, Director of the Agricultural Economics Research Institute in Oxford, gave a talk on the work of his Institute. Prior to 1924 the Institute was the only centre in Britain where formal provision had been made for research in the economic and financial aspects of landowning and farming. Before 1924 no serious attempt had been made, outside the University of Oxford, to give courses of lectures in agricultural economics. The Agricultural Economics Research Institute provided post-graduate training in the subject but the yearly number of students trained was small and insufficient to meet the needs of the agricultural economic advisory service. Consequently when Professor Ashby was appointed Head of the newly established Department of Agricultural Economics at U.C.W., Aberystwyth he organised courses of lectures in agricultural economics and rural sociology. By 1926, the university of Wales had approved a scheme of study leading to a joint Honours Degree in Economics with Agricultural Economics.

I decided to apply for a Ministry of Agriculture Senior Scholarship. But first I had to secure the Diploma in Economics and Political Science otherwise there was little hope that I would gain admission to a university. I was then in my twenty-sixth year. In my application for a scholarship I stated that I wished to go to U.C.W. Aberystwyth to study economics with agricultural economics. It was, at that time, the only place in Britain where one could study for a combined honours degree in these two subjects. Interviews for these scholarships took place in London. The interviewing committee included a representative from my union. This alarmed me, for if he held the same views as those expressed by Mrs Uzell I feared I could expect little support from him. When I met the committee I felt some members did not favour my application. I was asked whether, bearing mind the opportunities I had already had and the limited number of scholarships available, I did not consider I should make way for other applicants. It was no doubt difficult for the committee to distribute fairly ten senior scholarships over a large number of applicants. I was not sure how to answer the question. All I could do was remind them that in previous years applicants in a similar position to myself had been granteed scholarships. That being so I considered I should be given the same consideration.

Few applicants for these scholarships expressed a wish to study agricultural economics. This may have been an important factor in persuading the committee to give me one. For most students seeking admission to universities, at that time, agricultural economics was a little known subject. It was unlikely to attract students from towns or an industrial environment and most of those from rural areas, with an interest in agriculture, preferred to take courses based on the agricultural sciences.

At the time of applying to the Ministry of Agriculture for a scholarship I had made no formal request to U.C.W. for admission and the Ministry, when informing me that my application had been approved, stated that it was subject to my gaining admission to study agricultural economics at Aberystwyth. The grant was in the first instance for three years. I was advised to write to Professor Ashby. I did so and was invited to attend at his department for an interview. Having had no basic training in the physical and biological sciences I hoped to take, as Mr Hopkinson had done, the B.A. degree. After a long discussion Professor Ashby agreed to support my application for admission to the College, as a mature student, on the understanding that I studied agriculture as one of my two major subjects for the ordinary pass degree of B.Sc (Agriculture). The scheme of study for this degree included difficult first year courses in chemistry, botany, and one other science subject. The scheme was planned to allow students to choose any one of five alternative combinations of two subjects as major courses for the Honours degree. Professor Ashby insisted that some training in agriculture and the agricultural sciences was essential for agricultural economists working with farmers. The Ministry of Agriculture may have influenced this decision. Understandably it was considered important that students with Ministry of Agriculture scholarships should not pursue courses enabling them to take appointments outside agriculture and its ancillary industries. By insisting that their scholarship holders studied for an agricultural degree the risk of this happening was greatly reduced.

I was not sure that I would be accepted as a science student because of my lack of elementary training in chemistry, physics and the biological sciences. I was also doubtful about my chances of achieving, during the first three years, results which would persuade the Ministry of Agriculture to continue my grant for a fourth year. This was important to me since I wanted to work in the field of agricultural economics. In emphasising the difficulties I would have to face Professor Ashby increased my own doubts. He directed me to see Professor James, head of the chemistry department, who quetioned me closely on my previous education and in particular my training at Kirton Agricultural Institute. It was obvious he had serious doubts about my ability to cope successfully with the first year's work in chemistry. However, he gave me to understand that he would not oppose my application for admission. I now had the support of two professors but did not know if they would manage to persuade the heads of other science departments not to oppose my application. It was a great relief when I was informed that I had been accepted.

Male students unable to live at home had to find their own lodgings. This was not difficult. Aberystwyth was a popular seaside resort with plenty of accommodation for students during Michaelmas and Lent terms, when the town had few visitors. During the Summer term, especially after Whitsun, some students had to move into attic rooms, to share bedrooms and to take meals in dining rooms crowded with visitors. This upset occurred at a critical time when students were preparing for, and taking, their university examinations. For some students the only quiet place for studying in the evenings was away from holiday crowds on the top of Constitution Hill or some other place outside the town. I had arranged to co-dig with a Welsh ex-miner whom I met at the second of the two summer schools at Holybrook House in 1929. Like many miners, Jack Williams lost his work as a miner after the 1926 strike. In September, 1927 he gained a scholarship to Coleg Harlech, Merionethshire, which provided one year courses in non-vocational subjects. After spending three terms there Jack went to a Ministry of Labour training centre in Dudley where he learned to do joinery. While there he gained a W.E.A. scholarship to the Summer School and another from the same source which enabled him to go to U.C.W. Aberystwyth in October, 1929. When we met at Holybrook House we agreed to share digs in Aberystwyth. Later when we got to Aberystwyth we agreed to share living expenses.

Few landladies provided full board and lodgings. The most common practice was for students to pay a weekly rent, usually 10s., for their rooms and any necessary work in connection with the preparation of meals. Most landladies would provide a cooked meal for 1s. and our landlady did so on Sundays. On other days we either had cold lunches or took something in for the landlady to cook. We took weekly turns at doing the shopping. In this way we each had a chance to satisfy our particular likes. In order to minimise shopping we bought in a stock of sugar and jam at the beginning of each term. We became members of the local Co-op and decided to let our 'divi' rest there until the end of our training when we hoped to use it for a final celebration. Unfortunately the manager and our 'divi' disappeared in our third year. After that we shopped around for bargains.

Students who came to Aberystwyth after gaining the ordinary pass examinations at school could take an Honours degree in three years, the minimum period of residence for those taking intermediate degrees. Most students however, had, like myself, to spend at least four years to qualify for an Honours degree. In the first year I had to take three science subjects at the intermediate level. Chemistry and botany were obligatory first year subjects for agricultural students who could choose, as their third subject, geology, zoology or physics. I did not fancy cutting up dead animals and decided to take geology.

From my previous experience at Kirton Agricultural Institute and the warnings given by Professors Ashby and James I knew chemistry would prove difficult. This was confirmed as soon as I commenced the course of lectures and laboratory work. My fears about my ability to cope with the subject were not lessened when I learned, soon after my arrival in Aberystwyth, that one student, then in his third year, had, after several attempts, still not satisfied the examiners in first year chemistry. Like myself he had left school at an early age to work on farms. In 1926 he was given one of the Ministry of Agriculture junior scholarships and spent a year at Avoncroft Agricultural College. Then in 1927 he was awarded a senior scholarship and went to Aberystwyth to study for the B.Sc. degree in agriculture and the Honours degree in Economics with Agricultural Economics. Having had no previous training in the physical sciences he could not master the first year work in chemistry. After a number of failures it was feared he had become conditioned to a state of mind which made it impossible for him to produce satisfactory results in the examinations. However, he did so in his third year and subsequently left the university with a good second class Honours degree. When I learned of this student's experiences I was not surprised that Professor James had seemed uncertain about supporting my application to enter the College.

I was not the only student in that first year class with inadequate pre-university training in the sciences but was the only one without any basic training in chemistry and physics. A number had only an elementary knowledge of physics and in order to help them the lecturer in chemistry decided to give a short course of lectures on those aspects of physics of particular importance to his subject. I attended some of the lectures but soon discovered the subject was beyond my comprehension. During the first few weeks in Aberystwyth I lived in a world of make believe, trying to persuade myself I understood the lectures in chemistry and hoping that as we got further into the subject everything would become clear to me. As the first term progressed, however, I became more and more disturbed by my inability to understand the subject. The results of the terminal examinations at the end of the first term raised my hopes a little. I had gained reasonable marks in botany and geology and my mark in chemistry was not the lowest of the class. But after the examination in chemistry at the end of the second term I almost abandoned all hope of getting a university pass in the subject. From the start I knew I could not afford to take time off from my studies during vacations and from past experience knew it would be hopeless to try to study at home. I stayed in Aberystwyth during the Easter vacations, there I could work in the college library during the daytime and at my lodgings in the evenings. About a week before the Summer term, of my first year, I went along to the chemistry department to see how I had fared in the Easter examinations. In displaying the results names of students had been placed in four grades according to the marks gained. Hopefully I examined those in the second group and not finding my name there hoped it would be in the small group with top marks. As I well knew, it was not, neither was it in the third group with marks just above the failure level. My name was at the bottom of the failures with a mark of twenty-five per cent. As I turned away from the display board Professor James came out of his room. He asked how I was progressing with my work and how I had done in the examinations. I was so upset I could only point to the list. He took me into his room and asked about my difficulties. Before leaving he lent me a small book on chemistry which, I understood him to say, he had written for his small daughter. I must admit I had some difficulty with parts of this book but had not the courage to tell Professor James.

I commenced the summer term fearing the worst and as each week passed I became more and more apprehensive about my chances of satisfying the examiners. I felt reasonably certain of passing the examinations in botany and geology and therefore of being allowed to return for a second year. I went into the examinations in the summer term extremely nervous and in consequence did some foolish things, the most serious being my failure to deal with all the problems set in the geology practical test. I dealt with those on one side of the paper but failed to observe the instruction to turn the paper over for other problems. I finished what I thought was the complete paper in two hours, an hour short of the time allowed. This should have warned me to check my work and make certain I had not missed any part of the paper. I left the laboratory some time before the end of the three hours and as soon as I got outside discovered my mistake. It was too late, I was not allowed to return. There was no point in seeking an opportunity to explain to supervisors what had happened; the rules allowed no consideration for mistakes, genuine or otherwise. Examiners had no means of knowing whether I could have dealt successfully with the unfinished parts of the test. I left Aberystwyth for home fearing I had failed both geology and chemistry. It was, therefore, a great relief to learn I had satisfied the examiners in all three subjects.

In the next two years I studied agriculture and agricultural chemistry, agricultural botany, economic history, economics and agricultural economics. Having studied economics and economic history at Ruskin College I was told I need not attend lectures in these subjects during my second year but that I would have to take the university examinations at the end of the year. I attended some of the lectures in economics and all those given in economic history. It seemed to me advisable to know how lecturers at Aberystwyth treated economic theory and economic social problems. I enjoyed Professor Lewis's lectures in economic history; at each period he built up, on the blackboard, a synopsis of his treatment of the subject which I found very useful as a guide for further reading. At the end of the second year I passed the university examinations in agricultural botany, economics and economic history and at the end of the third year passed those in agriculture, agricultural chemistry, statistics and agricultural economics. In that year I gained a half share in the second prize offered for an essay on rationalisation of industry. I chose to write on the rationalisation of agriculture and was happy to know that in competition with Honours students taking economic science I had managed to gain a share of the second prize.

Towards the end of my third year I applied for an extension of my grant for a fourth year. The interview, in Shrewsbury, was immediately after the results of the examinations for the pass degree in agriculture had been published. As I had passed all my examinations at the first attempt I confidently expected to get an extension. I almost failed. I had been particularly anxious to get good results in economics and agricultural economics in order to be sure of being allowed to take these subjects in my Honours year. This caused me to give insufficient attention to agriculture which was, for me, the least interesting of the subjects taken. At the interview I gained the impression that my performance in agriculture and perhaps also in some of the other agricultural science subjects had not been outstanding. This, I understood, caused some members of the interviewing committee to doubt whether I was a suitable candidate for Honours courses. Fortunately my work in economics persuaded the committee to grant the extension.

In the university examinations for the Honours degree in

economics with agricultural economics, students had the choice of taking seven written papers, or five papers and presenting a dissertation on an approved subject. I was never satisfied with my performance in written examinations and elected to take the second option. It was agreed that my dissertation should be on Changes in the Agriculture of Lincolnshire since 1880. I spent the whole of the 1932 summer vacation collecting information and preparing the first draft of my thesis. This included a month in Aberystwyth extracting data from the Agricultural Statistics, published annually by the Ministry of Agriculture. The rest of the summer I spent at home extracting information from local papers and interviewing farmers. I worked on the project during the following Christmas vacation and by the middle of the Easter vacation had a final draft ready for typing. The thesis was more renowned for its quantity than its quality.

I was always in a nervous state when taking examinations and never felt confident of finding, at short notice, the right words to express my ideas. I went into the Honours examination in a very agitated state. In previous years I had fortified myself with a supply of chewing gum and managed the three-hour ordeal without having to ask to be taken to the toilets. In my fourth year chewing gum failed me on one occasion and it was rather embarassing being escorted by one of the invigilators the full length of a large hall, passing rows of students with smirks on their faces.

That year four other students took Honours in economics with agricultural economics and about a dozen took Honours in economic science. I was the only one who had elected to present a dissertation and had to be examined on it. My oral examination had been fixed for the morning after the meeting of examiners arranged to deal with the written papers. The results of the examination of all the Honours economics students, other than my own, were announced during the afternoon prior to my oral examination. Being, as usual, apprehensive about my own performance, I felt rather jealous of my fellow students who were in a carefree mood. In the evening I went to a dance organised for students leaving Aberystwyth after completion of their studies. During the evening a lecturer congratulated me on gaining a first class in my Honours examination. I was taken aback and explained that she must be mistaken as I had an oral examination on the following morning. She insisted that the documents had been signed. On the following morning I went along for my oral examination and Professor Ashby introduced me to the external examiner in agricultural economics who, shaking me by the hand, said he had read my thesis with great interest and had no questions to ask. He then congratulated me on my results.

Chapter 20.

UNIVERSITY DON

I THINK MOST OF MY FELLOW STUDENTS at Aberystwyth would have agreed that the town had some disadvantages as a centre for university education, the main one being its remoteness from important areas of population and of economic and commercial activities. The town, with a normal population of less than 10,000, was almost wholly concerned with servicing agriculture and catering for holiday makers. Agricultural students came mainly from farming stock. They had little opportunity, during their period of training, to get away from the influence of people primarily concerned, in one way or another, with agriculture; few chances to mix with workers and managers having wide experience of other industries and occupations. Had the college been in a large city, or better still in an industrial area, it is reasonable to assume that agricultural students would have gained a more balanced understanding of the importance of their industry in the economic and social life of Britain. As it was most of them gave the impression that their industry should not be subjected to the same economic tests as those applied to the manufacturing and servicing industries.

Sons and daughters of farmers, as students, resented criticism of farming systems and practices. They placed too much emphasis on unfair competition from imports, and on the industry's dependence on nature, as the cause of low incomes and low wages of farm workers. Few would admit that poor standards of management had been, to some extent, responsible for the industry's difficulties during the inter-war years. If they had been obliged to mix with people of wide industrial and commercial experience, they might have thrown off some of their false notions about the correctness of agricultural policies and practices. Only those students who took agricultural economics as one of their major courses spent any time studying the economic, social and political aspects of farming and rural communities.

Some may have thought that a student population of less than 1,000 was too small to provide, within the different disciplines, for economic use of staff and capital equipment. On the other hand the number was probably about right for a small country town. One wonders whether the relationship between local people and the present 3,000 students is as good as it was in the inter-war years. The number of students there 60 years ago, fitted into the life of a small town, heavily dependent on holiday trade. They did not put much money in the pockets of landladies but they made only minimal demands on local facilities for extra-curricular activities. There was never any noticeable disharmony between town and gown. There may be none today despite the trebling of student numbers, but residents may find it more difficult to accommodate themselves to the heavier demands made on local services despite a high proportion of students who now live in hostels. In my student days Welsh speaking students dominated college life. Today it is, I suppose, quite different with a very much larger proportion of students coming from homes outside the Principality.

Before going to Aberystwyth much of my small knowledge of Wales and its people was false. It was as unreliable as much of my knowledge of Scotland or many parts of England. This was not surprising nor peculiar to myself in view of the inadequacies of elementary education and the lack of opportunities for working class families to travel and learn by observation and personal contacts more about Britain. On immediate contact with students and townsfolk in Aberystwyth one was made aware that Wales was not a part of England, that Welsh people not only had their own language and culture but that many were determined to ensure that these should be given a more important place in the life of their country. I cannot remember having heard the Welsh language spoken before I went to Ruskin College. Until I went to Aberystwyth I did not know that in some parts of Wales little if any English was spoken, especially by the older people.

I soon learned that there was a growing demand by the young people of Wales for self government. As part of our studies on forms of government at Ruskin College I had examined various recommendations for some form of decentralisation of government. Some of my fellow students there supported the Independent Labour Party because it seemed more determined than the Labour Party to support self government for Wales and Scotland. The Labour Party in *Labour and the Nation*, published in 1928, stated the it 'would support the creation of separate legislative assemblies in Scotland, Wales and England with autonomous powers in matters of local concern.' For the people of Wales and Scotland it was important to know which particular areas of policy making and administration the Labour Party thought should be transferred to national legislative assemblies. To many people in Wales the declaration of intent seemed too vague; the language of politicians and of party documents could not be expected to satisfy ardent advocates of Welsh nationalism.

One advantage of the smaller number of students at Aberystwyth in my days there was that staff and students were known to each other. Lecturers were not anonymous persons lecturing to 300 or more students. It was, however, difficult to organise strong student societies. Agricultural students had an active literary and debating society which held six meetings in each of the Michaelmas and Lent terms. These were well attended by staff and students. The College Debating Union also held weekly meetings during the first two terms of each session. I enjoyed these but never acquired the skills and confidence of a good debater. I realise that Welsh people had a greater keeness for education than most people in my own country; that a much higher proportion of the children of ordinary people went to grammar schools and universities. In my own village very few sons and daughters of farm workers and of small farmers continued their education beyond the elementary schools.. As regards higher education I knew of no one, other than the sons of our vicar, who went to a university. In contrast any resident of a Welsh village of comparable size could at that time produce a long list of people who had been to, or were at, a university. Welsh parents took greater advantage of the grants provided by the Board of Education and local education authorities for young people wishing to go to training colleges and universities. They made greater personal sacrifices in favour of the education of their children. The fact that the then Board of Education contributed 60 per cent of the cost of education had a greater influence with Welsh than with English Local Education Committees. To the former it seemed a good investment to get £100 worth of education for only £40 on the local rates. In England, certainly in my own county, those in authority kept their eyes on the cost of education falling on local rates. It was said in the 1920s that our County Councillors boasted that we had the best roads and lowest costs of education in the country.

Apart from the local urge for education there was a strong economic reason why people in Wales should wish to educate themselves into professions. It was often said that lawyers, teachers and ministers represented the three most important exports from Wales. Because of the neglect of higher education in England it was, in the 1920s, much easier for Welshmen to obtain professional posts in England than in their own country. During the years of depression between the two world wars opportunities for economic and social advancement in Wales were very limited and in these circumstances it was sensible for young people of abililty to train themselves for professional posts in England.

The yearly number of young people in Wales seeking financial assistance to continue their education was much higher than could be fully met by local authorities during the years of financial stringency. Many students had to be content with small annual grants of £60, plus college or university fees, made by the Board of Education. There was a great deal of poverty among students, in some instances serious undernourishment. Some male students, from necessity, depended on chips for their mid-day meal and on fish and chips for an evening meal. Lack of a proper balanced diet was an important cause of tuberculosis among students at that time. In those days the price paid by some students for their education included a spell in one of the two sanatoria in Wales. It became so serious that in the 1930s the U. C. W. Aberystwyth authorities decided that every student in lodgings should have, at subsidised prices, a good substantial lunch on at least three days each week. Female students either lived at home or in college hostels; for them the problem of undernourishment was never so serious. When I learned that some students at Aberystwyth had to manage on their very small grants and loans I realised how fortunate I was to have a Ministry of Agriculture scholarship of £100 per year plus college fees and an initial grant of £40 for the purchase of clothes and other personal requirements.

During the 1930s many students in Britain demonstrated their opposition to war. In Aberystwyth we had, relative to our total numbers, a strong pacifist element who objected to the presence of the Officer Training Corp in our universities. In January, 1931, our Debating Union had before it the following motion:

That this House petitions His Majesty's Government to abolish Officers' Training Corps in the Universities.'

Two years later the Oxford Union voted in favour of a motion declaring that they would not fight for King and Country. Support for the motion debated at Aberystwyth came not only from students who loathed war but also from others who objected to the Civil Service Commission accepting lower standards in the Civil Service examination from students who had been members of the O.T.C. than those expected from other students. In their view there was no good reason for this preferential treatment. Many other students gave, in a variety of ways, service to the community and did so without monetary or other tangible rewards. It was not accepted that, in general, membership of the O.T.C. ensured that a student became a more valuable civil servant or member of society. There were two ways of getting rid of the practice, one by abolishing the O.T.C., the other by putting all candidates for the Civil Service Examinations on an equal footing. During the debate at Aberystwyth we feared that strong arm tactics would be used against those who spoke in favour of the motion during the open debate. Fortunately nothing more serious than pulling speakers down on to the floor occurred. The motion was carried by a show of hands but we knew this would not be accepted by the supporters of the O.T.C. They called for a poll of all students. I don't remember the results of the vote by ballot but it was of no importance for we knew our views would have no influence with the universities.

During February of each year there was a break in academic work for part of one week to allow the different sports clubs and cultural societies of the four constituent colleges of the University of Wales to compete in inter-college competitions. This was an occasion when Wales was a divided nation. There was keen rivalry between the colleges with much shouting, singing and waving of college scarves on the sports fields and in college halls. It was not possible to get all the athletic events organised into one week. Field competitions started a fortnight before 'Inter-Coll' week.

During the first two days of 'Inter-Coll ' week lectures continued in the mornings but for most students it was difficult to concentrate on work. There was too much excitement, too many evening engagements for any but the most serious student to have the inclination to attend lectures. The rest of the week was given over completely to competitions. Only those in their Honours year, determined to get good passes, could muster the will-power to continue working during the turmoil. Ardent enthusiasts of the competitions insisted that all students ought to attend the various competitive events and cheer on members of their own college to victory. They were highly critical of those who continued with their studies, but more particularly of those who took advantage of the break in lectures to go home for a few days. National and local Eisteddfodau occupy an important place in the cultural and social life of Wales and, as was to be expected, students of the four colleges included a one day Eisteddfod as one of the important events in 'Inter-Coll' week. Each college in turn provided facilities for this event and since most students spent four academic years in one or other of the Colleges each had an opportunity to compete, or listen to competitors, in at least one Eisteddfod.

The 1930s were not a promising time for finding suitable employment after the completion of one's studies. At that time few employers visited the universities to interview students during their final year of training. Universities, for their part, did not have on their staffs persons charged with the duty of assisting students to find suitable posts. Many students had to wait several months, in some instances more than a year, before obtaining appointments in posts requiring the special training they had received. Some, whose parents could finance them, returned to College to study for higher degrees, others took any kind of work that was on offer. The prospects for those seeking posts in agricultural economics was rather better than that for graduates in the agricultural sciences. A number of Marketing Boards established under the Agricultural Marketing Acts of 1931 and 1933 to deal with the marketing of farm produce, required a number of people with training in agricultural economics, in economics and statistics. It was also a time when the Ministry of Agriculture and the Agricultural Economics Advisory Service were increasing their staffs to cope with the extra work resulting from the agricultural policies being pursued at that time.

I did not make any applications for a post until after I had finished my studies. There was no encouragement from Professor Ashby to do so for he still had a few students from earlier years searching for posts. In the circumstances I was fortunate in having to wait only three months after leaving Averystwyth before my first appointment. This, like most of my subsequent appointments, was by invitation. The Head of the Department of Agricultural Economics, University of Reading, wrote offering me a post as student assistant in his department. At that time student assistants formed a high proportion of the small number of agricultural economists in the Agricultural Economics Advisory Service. I don't know what their status was at other universities but at Reading I had none of the privileges of the academic staff and was not eligible to be a member of the Senior Common Room. My salary of £150 was at that time subject to an economy cut of five percent. One had to be extremely careful to manage on a monthly cheque of just over £11. A car was necessary for my work and I was fortunate to have the £50 which I paid for my first second-hand car. I could not afford to run it for pleasure.

After eight months at Reading, Professor Ashby, in May 1934, invited me to return to Aberystwyth and join his staff as Permanent Technical Assistant at a salary of £250. This was a superannuated post. I was particularly grateful to Professor Ashby. He had been very helpful during my four years as a student and I felt he had already done more for me than I had a right to expect. He was recognised as the leading agricultural economist in this country and the scope for teaching work in agricultural economics was greater in his department than in any other British university outside Oxford.

Ten years earlier my vain hope had been that I might find some way of escaping from agriculture, from an industry which seemed to me to offer nothing but dreary work for low wages. It offered nothing which gave me the sense of being appreciated as a valued member of society. A farm labourer was considered a 'clod'. Little respect was shown him by the workers in factory, shop or office, by those who for so long had enjoyed the benefits of cheap food at his expense and that of his wife and children. I had not escaped from agriculture but had, at last, with the help of good fortune and kind friends, found work which gave me the opportunity to use whatever talents I had in a field of activity which for the rest of my working life gave me pleasure.

APPENDICES
Some associated letters and photographs.

> Mrs Pitkin, 12 Bushey-Hall-Road, Watford, Herts.
> Monday, April 14th, 1924
> To, Comrade J. H. Smith, Junior,
> In response to your appeal, re farm-lock-out at Sutterton, I have great pleasure in sending you 2/6, (unfortunately the utmost that I can run to,) toward just a very little relief for the farm-labourers' kiddies. Also I append the enclosed notice, (once advertised in the "Unemployed-Worker,) if you will please make it known, so that any Farm-Labourer's wife, possessing a baby, at Sutterton, or elsewhere, can have a free gift, (post-paid) of a baby's knitted hood, by applying, at the above address, to,
> Dear Comrade,
> Your's Fraternally,
> Annie Pitkin.

At the time of the agricultural workers' trade dispute (strike/lock-out) in 1924 – see Chapter 13 – the only financial help received from outside the Union (and little help was possible from within it) came from the local—labour—M.P. and from a Mrs Annie Pitkin, who had seen reports of the dispute in the Daily Herald. She lived in Watford and it would seem had poor regard for politicians (at least Conservative ones) and a large heart. Nothing more is known about her beyond these letters but they tell of a lady anxious to help those suffering from injustice although her means of helping were very limited. The first letter is reproduced in its entirety; the first part of her second indicates her proclivity for quoting from the Bible; and the beginning and end of her third letter show something of her opinion of politicians.

See above, opposite, and overleaf. Her handwriting is reproduced at approximately 60% size.

Mrs Pitkin, 12 Bushey-Hall-Road, Watford, Herts.
Tuesday, April 22nd, 1924.
To Mr J. H. Smith, (Junior,)
Dear Comrade,
        Enclosed a further 2/6 for my brave comrades on strike. I do so hope you will all be able to hold out right to the end, and so, win through at last. My P.C., re the Baby's Hoods, came back to me quickly. I sent off a hood almost at once. I am sending the card again to you, if at any time, you know of any body else, with a baby, who would care to receive one. Don't let the card itself be sent back to me, but, please keep it for a reference, then my address can be copied by any likely individual.

    Now, I am sending a message to your Strike-Committee. i.e. Will any, or all, of them, who possess Bibles, turn to the Epistle by James, chapter V, verses 1-6, inclusive. They will find these words:——
Go to now, Ye Rich Men, Weep and howl for your miseries that shall come upon you. Your riches be corrupted, and your garments are moth-eaten. Your gold and silver is cankered. The rust of them shall be a witness against you, and shall eat your flesh, as it were fire. Ye have gathered treasure together for the last days. Behold the hire of the labourers who have reaped down your fields, and who

Annie Pitkin's second letter showing Biblical references.

Mrs Pitkin, 12 Bushey-Hall-Road, Watford, Herts.
                Tuesday, May 12th, 1925
Dear Friend, and Comrade,
    I wonder if you can imagine how delighted I was to get your nice long letter this morning. I am so sorry that your brave efforts were all in vain. The mass of people still need a deal of educating upwards to get any tolerable result out of them. It is just the same here in Watford,

The beginning and end (omitting some three pages in the middle) of Annie Pitkin's third letter.

. . . . Now my last point is a tale one of the Devils was showing a man round Hell, he saw the murderers in one place, the thieves Clug as well as little, in another, and so on, burning away, at last he came to a cage where a lot were not burning at all. "Who are these? said the man. "Oh", the Devil replied "these are so green they won't burn properly yet; so we are drying them ready, these are the Conservatives" That's all.   Now, as I make it a solemn rule never to spend more than 4 pages on any one letter (or I should never be done,) I wind up, with my very real sympathy to you and your true comrades, on the part of
            Your's Fraternally,
                    Annie Pitkin.

206.

NATIONAL UNION of AGRICULTURAL WORKERS

Telephone MUSEUM 5015
Telegraphic Address NATAGRIC, KINCROSS, LONDON

AFFILIATED WITH TRADES UNION CONGRESS AND THE LABOUR PARTY.
APPROVED SOCIETY UNDER NATIONAL INSURANCE ACT REGISTERED Nº 393.
General Secretary,
R.B WALKER

Registered Office.
HEADLAND HOUSE,
308 GRAYS INN ROAD,
LONDON · W·C·1.

District Organiser
H. J. JONES.

"DUNHEVED," 8, WEST ROAD, BOURNE, LINCS.

25/4/24.

Dear Comrade,

Herewith I send you cheque value £10-16-0. being £7-16-0. due for Lock-out pay and £3 which I am lending for the supplementary fund, this amount we must raise in order that I am be repaid, In view of the difficult position, I again tender the advice which I gave when I came to Sutterton on the 6th inst, that all members should endeavour to obtain work on Farms where they will not be required to work, except by way of necessary overtime, on Saturday afternoons. I shall endeavour to meet you on Sunday at 2-30.P.M.

        Best Wishes,

        Your's sincerely,

        H. J. Jones

The help possible from the Union at the time of the 1924 strike/lock-out was, in practical terms, at best minimal: the moral support was perhaps not much greater.

> To all it may concern.
>
> This is to certify that Joseph H 11.11.37 Smith has been employed by me has a farm labourer on the farm of "Machin's Exors" Icsberton Bank. N.r Spalding Lincs for the past 12 months
>
> I have always found him trustworthy punctual & willing & can throughly reccomend him for any position as such
>
> Any further particulars will be gladly answered by me
>
> Yours Truly
> H. Bustey. "Foreman"
> Newlands Road
> Surfleet Seas End
> N.r Spalding
> Lincs

Whatever Joe thought of farm work it appears that he was found to be a thoroughly reliable and hardworking employee who may have believed that he and his companions were getting a raw deal but who nevertheless certainly gave his 'pound of flesh'.

208.

University College of Wales, Aberystwyth, Debating Society – probably taken in 1931.
Joe Smith is seated on the floor the second from the left.

SUTTERTON SUCCESS

Mr. J. H. Smith, of Marsh Road, Sutterton Dowdyke, who has recently taken his B.Sc. degree in Agricultural and Political Economics.

The only known photograph of Joe Smith in his cap and gown immediately after the degree ceremony at Aberystwyth when he received his B.Sc. degree.

The photograph is reproduced from a Lincolnshire newspaper but the cutting gives no indication as to which paper nor where nor when the photograph was taken. It was presumably a photograph taken in Aberystwyth and reproduced in his local newspaper.

Prifysgol Cymru. **University of Wales.**

Telephone Nos. {1776, 1777}

Replies should be addressed to the Registrar.

University Registry,
Cathays Park,
Cardiff.

June 24, 1957.

Dear Sir, or Madam,

    I am glad to be able to inform you that you have satisfied the examiners in the examination for the degree of M.Sc.

                  Yours faithfully,

                        Registrar.

Joseph Henry Smith, Esq., B.Sc.,
Bro Dawel,
Bow Street, S.O.,
Cards.

Four years after obtaining his First Class Honours degree Joe was awarded his Masters degree.
A significant achievement for a Lincolnshire plough-boy.